LONDON MATHEMATICAL SOCIETY LECTURE NOTE SERIES

Managing Editor: Professor J.W.S. Cassels, Department of Pure Mathematics and Mathematical Statistics, University of Cambridge, 16 Mill Lane, Cambridge CB2 1SB, England

The books in the series listed below are available from booksellers, or, in case of difficulty, from Cambridge University Press.

London Mathematical Society Lecture Note Series. 152

Oligomorphic Permutation Groups

Peter J. Cameron
School of Mathematical Sciences, Queen Mary and Westfield College

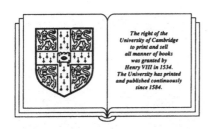

The right of the
University of Cambridge
to print and sell
all manner of books
was granted by
Henry VIII in 1534.
The University has printed
and published continuously
since 1584.

CAMBRIDGE UNIVERSITY PRESS

Cambridge

New York   Port Chester   Melbourne   Sydney

Published by the Press Syndicate of the University of Cambridge
The Pitt Building, Trumpington Street, Cambridge CB2 1RP
40 West 20th Street, New York, NY 10011, USA
10, Stamford Road, Oakleigh, Melbourne 3166, Australia

© Cambridge University Press 1990

First published 1990

Library of Congress cataloguing in publication data available

British Library cataloguing in publication data available

ISBN   0 521 38836 8

Transferred to digital printing 2001

Preface

A permutation group G on an infinite set Ω is said to be *oligomorphic* if G has only finitely many orbits in its induced action on Ω^n for all n.

The class of oligomorphic permutation groups has many links with other topics, notably model theory. According to the theorem of Engeler, Ryll-Nardzewski and Svenonius, a countable first-order structure has $\aleph_0$-categorical theory if and only its automorphism group is oligomorphic. Moreover, oligomorphic permutation groups have special features: properties such as primitivity and preservation of a linear order are first-order. We may consider only countable domains Ω without significant loss. We can with advantage replace such a group by its closure in the symmetric group (in the topology of pointwise convergence). Many examples can be constructed, and subgroup theorems proved, by the methods of Baire category and measure. Many sequences of integers of great combinatorial interest count orbits on n-sets or n-tuples in oligomorphic groups; indeed, calculating such sequences is equivalent to certain combinatorial enumeration problems.

These notes are based on lectures I gave at a LMS Durham symposium on "Model Theory and Groups" in the summer of 1988, and include contributions from a number of the participants. The notes were written in a great rush immediately after the Durham symposium (and I thank my family for their tolerance during our summer holiday). Though it has been considerably revised since then, I have tried to retain the informal style of the lectures Among many debts to them, I am grateful to Wilfrid Hodges and Dugald Macpherson for their detailed comments on the manuscript. I am also grateful to John Truss for saving me from making some rash conjectures (by disproving them); to colleagues at QMW (especially Francis Wright) and to Martyn Dryden (SWSL) for their help with TeX; and to Neill Cameron for the artwork. And finally, thanks to Roger Green for coining the unfamiliar word in the title.

I hope that these notes will encourage interest in a fascinating part of the rapidly vi

developing theory of infinite permutation groups.

Peter J. Cameron

School of Mathematical Sciences
Queen Mary and Westfield College
Mile End Road
London E1 4NS

January 1990

Contents

viii

1

Background

1. Background

1.1. HISTORY AND NOTATION

In the summer of 1988, a London Mathematical Society symposium was held in Durham on "Model Theory and Groups", organised by Wilfrid Hodges, Otto Kegel and Peter Neumann. This volume of lecture notes is based on the series of lectures I gave at the symposium, but is something more: since no Proceedings of the symposium was published, I have taken the opportunity to incorporate parts of the talks given by other participants, especially David Evans, Udi Hrushovski, Dugald Macpherson, Peter Neumann, Simon Thomas and Boris Zil'ber. (A talk by Richard Kaye revealed new horizons to me which I have not fully assimilated; but Richard's own book should appear soon.) In addition, I have made use of parts of the proceedings of the Oxford–QMC seminar on the same subject which ran weekly in 1987–8 and continues once a term (now as the Oxford–QMW seminar!); contributions by Samson Adeleke, Jacinta Covington, Angus Macintyre and John Truss have been especially valuable to me.

Why model theory and groups? In particular, why the special class of permutation groups considered here?

In the middle 1970s, when my interests were entirely finite, John McDermott asked a question about the relationship between transitivity on ordered and unordered n-tuples for infinite permutation groups. The analogous question, and more besides, had been settled for finite permutation groups by Livingstone and Wagner (1965), with techniques which were largely combinatorial and representation-theoretic, and so not likely to be useful here. McDermott himself had constructed some examples showing that the infinite is very different from the finite.

At that time, "infinite permutation groups" could scarcely be described as a subject. In the Mathematical Reviews classification, permutation groups were explicitly finite.

The only work of substance was Wielandt's Tübingen lecture notes (1959), which was not readily available. Moreover, of the few results which were in the literature, a substantial proportion were by topologists (such as Anderson (1958) and Brown (1959)), and relatively unknown to group theorists.

Another symptom of the situation is illustrated by the following three theorems.

Tits (1952): There is no infinite 4-transitive group in which the stabiliser of 4 points is trivial.

Hall (1954): There is no infinite 4-transitive group in which the stabiliser of 4 points is finite of odd order.

Yoshizawa (1979): There is no infinite 4-transitive group in which the stabiliser of 4 points is finite.

Yoshizawa's theorem is not so much harder than the other two; why, then, the quarter-century gap? In the first two cases, the theorems were regarded as little more than footnotes to the complete determination of finite permutation groups with the same properties (viz. small symmetric, alternating and Mathieu groups). A finite version of Yoshizawa's theorem would be a list of all 4-transitive groups. This was out of reach in the 1950s and 1960s, and by the 1970s it was clear that it would be obtained as a corollary of the classification of finite simple groups; this duly happened in 1980. But the imminence of this classification had also made people think that interesting problems on infinite permutation groups might be waiting.

In addition, there was pressure from outside, especially from model theory. Questions that arise naturally in classification theory and enumeration of models lead to problems about structures with large automorphism groups. Work of Fraïssé and his school (notably Frasnay and Pouzet) leads in the same direction. (See Fraïssé (1986).) Another contribution was from Joyal (1981), who was developing a subject which included "Redfield-Pólya enumeration without groups".

To return to my personal narrative. I was able to answer John McDermott's question and give a classification of permutation groups which are transitive on unordered n-sets for all n. I spoke about this in Oxford, and in the pub afterwards Graham Higman said, "What about groups with finitely many orbits on n-sets for all n? That might be a good topic for a research student." I have researched and studied this topic on-and-off since then, and now I present my thesis.

A permutation group on an infinite set is called oligomorphic if it satisfies the condition of Higman's question (or, equivalently, if it has only finitely many orbits on n-tuples for all n). The main connections between oligomorphic permutation groups and the areas of model theory and combinatorics are provided by two key results which will be described further in Chapter 2, but which can be stated loosely as follows:

Ryll-Nardzewski's Theorem: A countable (first-order) structure is axiomatisable (that is, characterised, up to isomorphism, as a countable structure, by first-order sentences) if and only if its automorphism group is oligomorphic.

Fraïssé's Theorem: The problem of calculating the numbers of orbits, on n-sets or on n-tuples, of oligomorphic permutation groups is equivalent to that of enumerating the unlabelled or labelled structures in certain classes of finite structures (characterised, more-or-less, by the amalgamation property).

These two results provide the central theme of my lecture notes.

The remainder of this chapter tries to provide a crash course in some of the techniques needed later: first, the two areas principally involved, namely permutation groups and model theory; then, two areas which provide important tools, namely category and measure, and Ramsey theory. The relevant sections can safely be skipped by an expert. Chapter 2 presents the basic properties of oligomorphic permutation groups, and their connection with the theorems of Ryll-Nardzewski and Fraïssé. The third chapter discusses properties of the sequences enumerating orbits on n-sets or n-tuples, especially their growth rates. In Chapter 4 I turn to subgroup theorems explaining how techniques of measure and category, combined with Fraïssé's theorem, allow us to construct various interesting subgroups of closed oligomorphic groups. One of the theorems here has an application to the theory of measurement in mathematical psychology! The final chapter treats some important but miscellaneous topics.

For further reading on the topics of Chapter 1, see Wielandt (1964) for permutation groups, Chang and Keisler (1973) for model theory, Oxtoby (1980) for measure and category, and Graham, Rothschild and Spencer (1980) for Ramsey theory. Other useful books in related areas are those of Fraïssé (1986), Goulden and Jackson (1983) and Shelah (1978).

The exercises are a mixed bag, and should *not* be regarded as routine tests of comprehension. Some are very difficult, and I have given hints, of which the most detailed are outline proofs. Unsolved problems slipped in among the exercises are flagged as such; others are scattered through the text.

A few comments about terminology. As in logic, the natural numbers begin at 0. But I am not totally consistent: I use $\mathbb{N}$ rather than ω to denote the natural numbers, and $\aleph_0$ for their cardinality; and if I want to refer to just two objects, I usually number them 1 and 2 rather than 0 and 1. However, I use the term "ω-sequence" for an infinite sequence (of order type ω). Also, unlike the logicians, I don't insist that the domain of a structure be non-empty. As in model theory, it is convenient to treat n-tuples flexibly, regarding them as "ordered n-subsets". Thus, if I say that n-tuples $\bar{a} = (a_1, \ldots, a_n)$ and $\bar{b} = (b_1, \ldots, b_n)$ carry isomorphic substructures, I mean that the map $a_i \mapsto b_i$ $(i = 1, \ldots, n)$ is an isomorphism between the induced substructures on $\{a_1, \ldots, a_n\}$ and $\{b_1, \ldots, b_n\}$.

As usual, $\mathbf{Z}$, $\mathbf{Q}$, and $\mathbf{R}$ are the integers, rationals and real numbers.

1.2. PERMUTATION GROUPS

A *permutation group* G on a set Ω is simply a subgroup of the symmetric group on Ω (the group of all permutations of Ω). However, to allow us to consider a number of permutation groups isomorphic to (or homomorphic images of) a fixed group G, a more general concept is convenient. A *permutation representation* of G on Ω is a homomorphism from G to the symmetric group on Ω. Other terminology is often used: we say that G acts on Ω, or that Ω is a G-set or G-space. The image of the homomorphism is a permutation group, denoted G^Ω, and called the permutation group on Ω *induced* by G.

A permutation representation of G on Ω can be described by a function $\mu : \Omega \times G \to \Omega$, where $\mu(\alpha, g)$ is the image of α under the permutation corresponding to g. This function satisfies
  (a) $\mu(\alpha, gh) = \mu(\mu(\alpha, g), h)$;
  (b) $\mu(\alpha, 1) = \alpha$.
(These are translations of the closure and identity axioms for a group. Note that the other two group axioms do not have to be translated — composition of permutations is always associative; and the condition derived from the inverses axiom, namely
  (c) $\mu(\alpha, g) = \beta \iff \mu(\beta, g^{-1}) = \alpha$,
is a consequence of (a) and (b).) Conversely, given a map μ satisfying (a) and (b), the function carrying g to the permutation $\alpha \mapsto \mu(\alpha, g)$ is a permutation representation of G.

I will from now on suppress the function μ and write αg for the image of α under g. (Notice that I have sneaked in the convention that permutations act on the right!)

Let G act on Ω. Set $\alpha \sim \beta$ if there exists $g \in G$ with $\alpha g = \beta$. This is an equivalence relation on Ω; the reflexive, symmetric and transitive laws correspond naturally to conditions (b), (c) and (a) above. Its equivalence classes are called *orbits*, and G is called *transitive* if it has but one orbit. If a subset Δ of Ω is a union of orbits, then we have an action of G on Δ. In the case when Δ is a single orbit, the permutation group G^Δ induced on Δ is called a *transitive constituent* of G.

The transitive constituents of a permutation group do not determine it uniquely; but we have:

(1.1) *Any permutation group is a subgroup of the cartesian product of its transitive constituents.*

The cartesian product of $(G_i : i \in I)$ is the set of functions $f : I \to \bigcup G_i$ such that $f(i) \in G_i$ for all $i \in I$; the group operation is componentwise. To each $g \in G$ corresponds the function f_g for which $f_g(i)$ is the restriction of g to the ith orbit: this defines the embedding of G in the cartesian product. In fact, G is a *subcartesian product* of its transitive constituents. (This simply means that it projects onto each factor of the product.)

Note that we have the cartesian product here rather than the (restricted) direct product (which consists of those functions f for which $f(i) = 1$ for all but finitely many $i \in I$). Of course, if there are only finitely many orbits, then the two are indistinguishable.

Given any family $(G_i : i \in I)$ of transitive permutation groups, their cartesian product has a natural action for which the G_i are the transitive constituents. When I refer to the cartesian (or direct) product of permutation groups, this action is intended. There are other actions, which will sometimes be needed; for example, there is an action on the cartesian product of the domains (rather than the disjoint union).

Let G act on Ω. The *stabiliser* G_α of a point $\alpha \in \Omega$ is the set $\{g \in G : \alpha g = \alpha\}$. It is a subgroup of G. Similarly, if $\Delta \subseteq \Omega$, the setwise stabiliser G_Δ of Δ consists of all permutations $g \in G$ which map Δ onto itself; and the pointwise stabiliser $G_{(\Delta)}$ is the set of permutations which fix every point of Δ. Often, $\bar\alpha$ will denote an ordered tuple of elements of Ω, and then $G_{\bar\alpha}$ will denote the pointwise stabiliser of $\bar\alpha$.

Let H be a subgroup of the abstract group G. The *coset space* of H in G is the set of right cosets of H in G; there is an action of G on it given by $(Hx)g = Hxg$ (or, more pedantically, $\mu(Hx, g) = Hxg$). Coset spaces provide "canonical" transitive G-spaces:

(1.2) *If G acts transitively on Ω, then Ω is isomorphic to the coset space of G_α in G, for $\alpha \in \Omega$.*

(A *G-isomorphism* between G-spaces Ω_1, Ω_2 is a bijection θ such that, for all $g \in G$, the diagram

$$
\begin{array}{ccc}
\Omega_1 & \xrightarrow{\ \theta\ } & \Omega_2 \\
\downarrow{\scriptstyle g} & & \downarrow{\scriptstyle g} \\
\Omega_1 & \xrightarrow{\ \theta\ } & \Omega_2
\end{array}
$$

commutes.)

G acts *regularly* on Ω if it is transitive and the stabiliser of a point is the identity. By (1.2), in this case, Ω is isomorphic to G (on which G acts by right multiplication) — this is called the *right regular representation*. In this situation, G acts *faithfully* on Ω; that is, the map taking elements of G to the corresponding permutations is one-to-one, so that G is isomorphic to its image. This action was used by Cayley to show that every group is isomorphic to a permutation group.

Now suppose that G is transitive on Ω. A *congruence* is a G-invariant equivalence relation on Ω. There are two trivial congruences, namely, equality and the "universal" relation with a single equivalence class. G is said to be *primitive* if there are no other congruences. A transitive group G is primitive if and only if G_α is a maximal subgroup of G. (More generally, the congruences form a lattice isomorphic to the lattice of subgroups lying between G_α and G.)

Once again, if G is not primitive, we can break it down into "smaller" pieces. The reverse construction is the wreath product of permutation groups, defined as follows. Let H and K be permutation groups on Γ and Δ respectively. Take $\Omega = \Gamma \times \Delta$, thought of as a covering space of Δ with fibres bijective with Γ. (See Fig. 1.) Let B (the base group) be the cartesian product of $|\Delta|$ copies of H, one associated with each fibre of Ω (that is, each element of Δ); and let K_1 (the top group) be the permutation group on Ω obtained by letting K permute the fibres according to its given action on Δ. Then the *wreath product* $H \operatorname{Wr} K$ is the (semi-direct) product of B and K_1.

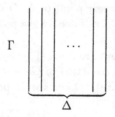

Fig. 1. A covering space

(1.3) *Let G be transitive but imprimitive on Ω. Let Γ be a congruence class; H, the permutation group induced on Γ by its setwise stabiliser in G; Δ, the set of congruence classes; and K, the group induced on Δ by G. Then G can be embedded in a natural way into $H \operatorname{Wr} K$ (as permutation group).*

There is a more general definition of wreath products, which can take account of an arbitrary (partially ordered) set of congruences; but I shall not require this.

Any finite transitive permutation group can be analysed or "broken down" into primitive "components" in finitely many steps (though of course some information is lost). This is not true in general for infinite permutation groups; but it is the case for the groups I shall be considering, those with only finitely many orbits on n-tuples for all n (or indeed, just for $n = 2$). This is because such a group can have only finitely many congruences. (Any congruence, thought of as a set of ordered pairs, is a union of orbits of G in its action on $\Omega \times \Omega$.)

Another important action of the wreath product $H \operatorname{Wr} K$ is the *product action*, on the set of functions $\phi : \Delta \to \Gamma$. (An element f of the base group acts by

$$\phi \cdot f(i) = \phi(i) \cdot f(i),$$

while the top group acts on the arguments of the functions by

$$\phi \cdot k(i) = \phi(ik^{-1}).$$

Exercises

1. For $H \leq G$, the kernel of the action of G on the coset space of H is $\bigcap_{g \in G} g^{-1} H g$. (This subgroup, called the *core* of H in G, is the largest normal subgroup of G which is contained in H.)

2. The only primitive regular groups are the cyclic groups of (finite) prime order.

3. Let H, K act on Γ, Δ respectively. If $|\Gamma|, |\Delta| > 1$, show that $H \operatorname{Wr} K$ (in its product action) is primitive if and only if H is primitive but not regular on Γ and K is transitive on Δ.

4. Prove the assertion in the text that, if G is transitive on Ω, then the lattice of congruences is isomorphic to the lattice of subgroups between G_α and G. (Consider the stabilisers of the congruence classes containing α.)

1.3. MODEL THEORY

Model theory concerns the relationship between sentences in a formal language and the structures satisfying them. The language appropriate here is that of first-order logic.

The common features of first-order languages are the logical connectives ($\neg$ (*not*) and $\rightarrow$ (*implies*) suffice, though we also use $\vee$ (*or*) and $\wedge$ (*and*)), and quantifiers ($\forall$ (*for all*) and $\exists$ (*there exists*)), punctuation marks (parentheses and comma), and a supply of variables (countably many will be enough). In addition, a language contains symbols for functions, relations and constants, appropriate to the application (the area of mathematics being modelled). Each relation or function symbol is equipped with an *arity* (the number of arguments it takes) as part of its syntax. The language. is called *relational* if it contains no function or constant symbols.

For example, for group theory, we could take a binary function (for multiplication), a unary function (for inversion), and a constant (the identity); or we could make do with the first of these alone; or we could define multiplication by a ternary relation R, so that $R(x, y, z)$ is interpreted to mean $xy = z$.

Formulae are defined recursively in a standard way. First, a term is a constant symbol or a variable or a function symbol with the correct number of terms as arguments. An atomic formula is a relation symbol with terms as arguments; a general formula is obtained by combining formulae by means of connectives, or preceding them with quantifiers. It is a sentence if it has no free (unquantified) variables.

A sentence is *universal* ($\forall$), *existential* ($\exists$), or *universal-existential* ($\forall\exists$) if it has the form $(\forall \bar{x})\phi(\bar{x})$, $(\exists \bar{x})\phi(\bar{x})$, or $(\forall \bar{x})(\exists \bar{y})\phi(\bar{x}, \bar{y})$ respectively, where ϕ is quantifier-free.

A structure over a language consists of a set equipped with distinguished constants (i.e. elements of the set), functions and relations corresponding to the symbols in the language (and having the appropriate arities). It is hopefully clear what it means for a sentence ϕ to be satisfied, or valid, in a structure M: we write $M \models \phi$ and say that M is a *model* of ϕ in this case. Similarly for sets of sentences. Thus, a group is a model for the axioms of group theory.

There is a formal deduction system associated with a first-order language. Thias consists of a set of sentences called axioms, and some rules of inference which allow sentences to be derived from others (possibly in the presence of sets of "hypotheses"). In fact, there are several such systems; but all standard ones satisfy *Gödel's completeness theorem*:

(1.4) *A set of sentences has a model if and only if it is consistent (that is, no contradiction can be derived from it).*

From this major result, the two "portals" of model theory are derived.

(1.5) (The compactness theorem.) *A set of sentences has a model if and only if every finite subset has a model*

For, by (1.4), we can replace "has a model" by "is consistent"; and, since proofs in the formal system are finite, if a contradiction could be deduced, then only finitely many hypotheses would be used in the deduction, and this finite set would be inconsistent.

(1.6) (The downward Löwenheim-Skolem theorem.) *A set of sentences over a countable language which has a model has a finite or countable model.*

This comes from the proof of (1.4): the model constructed in that proof is countable. (See the note about equality below.)

Observe that (1.5) and (1.6) contain no reference to the deduction system. Indeed, it will play no further rôle in the discussion.

Equality is obviously important enough to have its own name and conventions. It can be shown, using the compactness theorem, that no set of axioms can force the interpretation of a binary relation to be equality; we can only say that it is an equivalence relation such that "equal" terms can be interchanged in formulae without changing their truth. Then, by factoring out the equivalence relation, we obtain a new structure which satisfies exactly the same sentences as the old one, in which the binary relation really is interpreted as equality. (Such a model is called *normal.*) Because of the importance of equality, it is customary to consider only normal models, and I shall follow this convention. Note, however, that the countable model constructed in the proof of (1.4) may not be normal, and the model obtained by "normalising" it may be finite; hence the possibility of a finite model in (1.6).

The compactness theorem is the source of many results about the limitations to what can be said using first-order sentences. Obviously, we can say everything about a finite structure, simply by listing all the instances and non-instances of relations, etc. But, for example, there is no set Σ of sentences such that all models of Σ are finite but their cardinalities are unbounded. For let ϕ_n be the sentence saying "there exist at least n points". (See Exercise 1: ϕ_2 is $(\exists x_1)(\exists x_2)(\neg(x_1 = x_2))$.) If such a Σ existed, then Σ together with any finite set of the sentences ϕ_n would have a model, but Σ together with all the ϕ_n would not.

Along the same lines, we have:

(1.7) (The upward Löwenheim-Skolem theorem.) *If a set of sentences has an infinite model, then it has arbitrarily large infinite models.*

To see this, we adjoin to the language a large infinite set of new constant symbols c_i, and to the set of sentences all those of the form $c_i \neq c_j$ (for $i \neq j$). Any finite subset is satisfiable (in the given infinite model), so the whole set is.

We use the term "theory" for "consistent (or satisfiable) set of sentences". There are two opposed points of view here. Some theories, like that of groups, are intended to have many different models; a logical consequence of the theory will be valid in all of them. Others are intended, as far as possible, to describe a single structure. A theory Σ is said to be *complete* if, for every sentence ϕ, either ϕ or $\neg\phi$ is in Σ. This is equivalent to saying that Σ consists of all sentences which hold in some fixed structure M. (We speak of the *theory of M*, written $\text{Th}(M)$.) If M is infinite, then Σ does not determine M, even up to cardinality (by (1.7)). The best we can expect is that M is the only model of Σ of its cardinality. This concept, for countable M, will be very important to us (see §2.5).

Exercises

1. Write down sentences ϕ_n, ψ_n (in a language with equality) such that
   (a) any model of ϕ_n has at least n elements;
   (b) any (normal) model of ψ_r has exactly n elements.

2. Use the compactness theorem to show that a theory (in the language of graphs) having models with arbitrarily large finite diameter has a model with infinite diameter.

3. Show that M is the unique model of $\text{Th}(M)$ (up to isomorphism) if and only if M is finite.

4. Write down a sentence, using equality and one binary relation symbol, all of whose models are infinite. Is this possible with equality alone?

1.4. CATEGORY AND MEASURE

Cantor's celebrated proof of the existence of transcendental numbers went like this: there are so many more complex numbers than algebraic numbers, that there are as many transcendental numbers as complex numbers. The related techniques of Baire category and measure allow refinements on this argument: certain subsets are "small", even though their cardinality is the same as that of the whole set.

Let (X, d) be a complete metric space. A subset of X is dense if its closure is X, i.e. if it meets every open set. A subset is *residual* if it contains the intersection of countably many open dense sets. (Other terms are used: the complement of a residual set is called *meagre*, or *of the first category* — whence a residual set is called *comeagre* — and a set which is not of the first category is *of the second category*.)

   (1.8) (The Baire category theorem.) *A residual set in a complete metric space is non-empty.*

Thus, if we can show that the set of elements having some property P is residual, then it follows that some element has the property P. But the interpretation of (1.8) is that a residual set is, in a sense, "large", containing "almost all" of the space; for example, it meets every open dense set. Also, the intersection of countably many residual sets is residual (and hence non-empty).

Metric spaces in these notes always arise in the following way. A point of the space is determined by a countable sequence of choices, and the nearness of two points depends on the initial segment of the choice sequences determining them which agree; we can take

$$d(x, y) = \frac{1}{2^n}$$

if the choice sequences for x and y differ first in the nth term. (The actual value we choose for the distance is not crucial; any decreasing function of n will do. The particular choice is motivated by consideration of Hausdorff measure, see Cameron (1987a).)

For an illustrative example, consider the case where there are just two alternatives for each choice, so that points are represented by an ω-sequence of zeros and ones. (Think of the outcome of countably many coin tosses as determining a point.) Now an open ball consists of all zero-one sequences with a given initial segment. Hence a set S is open if every point of S has an initial segment, all of whose continuations lie in S; and S is dense if every finite sequence has a continuation which lies in S. For this metric space, the Baire category theorem is easily proved directly (see Exercise 3).

This metric space is homeomorphic to the Cantor ternary set, the set of real numbers in the unit interval whose ternary expansions contain only the digits 0 and 2. (Replace all the ones in a given zero-one sequence by twos, and regard the result as the ternary expansion of a real number.) Note that, if we simply regard the given sequence as the binary expansion of a real number, the resulting map is not a homeomorphism, since it fails to be one-to-one.

The procedure for constructing a measure space is more technical, though in many cases there is an intuitive description. The general technique parallels the construction of Lebesgue measure on $\mathbf{R}$. First we define the measure on a basic class of sets (the open intervals in $\mathbf{R}$); then we extend it by countable additivity to the σ-algebra they generate (the Borel sets in $\mathbf{R}$); then finally (though this is not necessary in many applications) we define inner and outer measure on any subset, and call a set measureable if its inner and outer measure coincide.

In the case of the space of zero-one sequences, the basic sets are those with prescribed initial segment (that is, the open balls), and the measure of the set of sequences with a given initial segment of length n is $1/2^n$. There are two other ways to view this:

(a) We regard a sequence as giving the outcome of infinitely many tosses of a fair coin (so that the outcome of each toss has probability $\frac{1}{2}$, and different tosses are independent). Then we have the standard probability measure.

(b) Identifying a sequence with the binary expansion of a real number in the unit interval, we use Lebesgue measure on the interval. As we noted earlier, this map is not one-to-one; but the failure is not damaging here. (There is a countable, and hence null, set of reals which have two pre-images each.)

The "large" sets in this context are those of measure 1. As with residual sets, they have the property that the intersection of countably many such sets is again of measure 1, and hence non-empty. (This is familiar in the complementary form: the union of countably many null sets is a null set.) Moreover, any non-empty open set has positive measure; so a set of measure 1, like a residual set, is dense.

In the more general case where a point is determined by a sequence of choices which are not restricted to two alternatives each, the assignment of measure to open balls is not so straightforward. The measure of a given open ball has to be split up among its immediate successors. Equal division may not be appropriate. We will see examples later.

Exercises

(All these exercises refer to the space of zero-one sequences.)

1. Prove that, with the metric described in the text, the space of zero-one sequences is a metric space satisfying the *ultrametric inequality*, viz.

$$d(x, z) \leq \max(d(x, y),\ d(y, z)).$$

Show that, in an ultrametric space (one satisfying the ultrametric inequality),
  (a) every point in an open ball is its centre;
  (b) if two open balls intersect, then one contains the other.

2. Prove that the space of zero-one sequences is a complete metric space, and that its topology is that of pointwise convergence.

3. Show that a residual set is dense and has cardinality $2^{\aleph_0}$.

[*Hint.* Let S be residual and let $S = \bigcap\{X_n : n \in \mathbf{N}\}$, where each X_n is open and dense. To prove that $S \neq \emptyset$, define finite sequences σ_n inductively as follows:

  σ_0 is the empty sequence;
  if σ_n is defined, choose an infinite continuation s_n of it lying in X_n (possible since X_n is dense) and a finite initial segment σ_{n+1} of s_n all of whose continuations lie in X_n (possible since X_n is open) such that σ_{n+1} is longer than σ_n. The "limit" of the sequences σ_n lies in S.

Now, if U is a given open set, modify the construction by choosing σ_0 so that all continuations of σ_0 lie in U. To demonstrate the cardinality, "code" infinite zero-one sequences into the construction by adding one extra bit to each σ_{n+1}.]

4. A sequence is called *universal* if it contains every finite zero-one sequence as a consecutive subsequence. Show that the set of all universal sequences is residual and has measure 1, i.e. is "large" in both senses.

5.Show that the set of sequences with upper density 1 and lower density 0 is residual.

[The *upper density* of a zero-one sequence s is defined as

$$\limsup_{n\to\infty} d(n)/n,$$

where $d(n)$ is the number of ones among the first n terms of s; the *lower density* is the lim inf of the same quantity.]

Remark. By contrast, the strong law of large numbers asserts that the set of sequences having density 1/2 has measure 1. So, in this instance, the two techniques give conflicting views of the "typical" set. We will see other examples later.

1.5. RAMSEY'S THEOREM

The "pigeonhole principle" asserts that, if the infinite set X is partitioned into finitely many parts, then one at least of these parts is infinite. Ramsey's theorem is a generalisation of this.

   (1.9) *Suppose that the set of n-element subsets of the infinite set X is partitioned into finitely many parts. Then there is an infinite subset Y of X, all of whose n-element subsets belong to the same part of the partition.*

I'll prove this for $n = 2$, deducing it from the pigeonhole principle. The general proof is by induction on n, following the same lines, and is outlined in the Exercises.

Let $\{x_0, x_1, \ldots\}$ be an infinite subset of X. We choose an infinite subsequence of (x_i) as follows. Set $y_0 = x_0$. After the $(i-1)^{\text{st}}$ stage, $y_0, \ldots, y_{i-1}$ have been chosen, and there are infinite subsets $Y_1, \ldots, Y_{i-1}$ such that, for $j < i$,
   (a) $Y_j \subseteq Y_{j-1}$;
   (b) $y_j = \min(Y_j)$;
   (c) all edges from y_{j-1} to Y_j (that is, all 2-sets $\{y_{j-1}, z\}$ for $z \in Y_j$) lie in the same part of the partition.

In the ith stage, partition $Y_{i-1} \setminus \{y_{i-1}\}$ so that x lies in the kth part of the partition if and only if $\{y_{i-1}, x\}$ lies in the kth part of the original partition; let Y_i be an infinite part (guaranteed by the pigeonhole principle), and $y_i = \min(Y_i)$.

Now, in the subsequence (y_i), the number m_i of the part of the partition containing a pair $\{y_i, y_j\}$ $(i < j)$ depends only on i, not on j. Another application of the pigeonhole principle yields a subsequence on which m_i is constant. This is the required infinite set.

It is usual to express Ramsey's theorem in the language of colours and colourings. Associating a colour with each part of the given partition, we are provided with a colouring of the n-element subsets of X with finitely many colours. A subset of X is called "monochromatic" if all its n-element subsets have the same colour. Then the theorem asserts the existence of an infinite monochromatic subset of X.

Ramsey's theorem stands at the origin of a flourishing subject, some of whose concerns are quantification of the infinities involved, finite analogues, and similar results for structures other than sets. All I need, however, is a single generalisation of Ramsey's theorem:

   (1.10) *Suppose that the n-subsets of an infinite set X are coloured with r colours, all of which are used. Then there are an ordering $c_1, \ldots, c_r$ of the colours, and infinite sets $X_1, \ldots, X_r$, such that X_i contains a set of colour c_i but no set of colour c_j for $j > i$.*

Ramsey's theorem gives us a colour c_1 and an infinite set X_1. We proceed to "rank" the colours; a colour is ranked once we have found an infinite set containing n-sets of that colour and previously ranked colours only. The theorem is proved when all colours have been ranked; and Ramsey's theorem allows us to rank the colour c_1.

Suppose that, at some stage, c is a colour not already ranked, and C an n-set of colour c. The subsets of C are partially ordered by inclusion, and this partial order can be extended to a total order

$$\emptyset = C_0 \prec \ldots \prec C_s = C,$$

where $s = 2^n - 1$. Let Y be an infinite set disjoint from C and containing only sets of ranked colours.

We now proceed through the sequence (C_i), defining infinite sets Y_i, starting with $Y_0 = Y$. At the ith stage, we define a new colouring of the $(n - |C_i|)$-subsets of Y_i, by giving each such set B the colour originally assigned to $B \cup C_i$. There is an infinite set Y_{i+1} with all of its subsets having the same colour. Then all colours of n-sets occurring within $Y_{i+1} \cup C_i$ have been ranked except possibly for the unique colour of n-sets containing C_i; so this colour can be ranked if it hasn't already been. By the time we have worked our way through the entire sequence, the colour of C will have been ranked.

Now just continue this process until all colours are ranked and the theorem is proved.

Exercises

1. Prove Ramsey's theorem.

[*Hint:* The proof is by induction on the size n of the subsets being coloured. The start $n = 1$ of the induction is the pigeonhole principle, and the argument given illustrates the step from 1 to 2. In general, replace the first (but not the second) application of the pigeonhole principle with the inductive hypothesis for $n - 1$.]

2. The finite form of Ramsey's theorem is the following assertion:

Let n, m, r be given positive integers, with $n < m$. Then there is an integer N (depending on n, m, r) with the following property:

If the n-element subsets of an N-element set X are coloured with r colours, then there is an m-element subset Y of X, all of whose n-element subsets have the same colour.

Prove this, by a modification of the argument outlined above.

Harder: Deduce it from the infinite form of Ramsey's theorem by means of the Compactness theorem.

3. Formulate and prove a finite form of (1.10).

4. Show that an infinite sequence of elements of a totally order set contains one of the following: a constant subsequence; a strictly increasing subsequence; a strictly decreasing subsequence.

Using this fact, deduce the Bolzano-Weierstrass theorem (for **R**) from the "principle of the supremum".

2

Preliminaries

2. Preliminaries

2.1. THE OBJECTS OF STUDY

In these notes, I will be concerned with the following situation:

G is a group of permutations of an infinite set Ω with the property that, for every natural number n, G has only a finite number of orbits on Ω^n, the set of n-tuples of elements of Ω.

Such a permutation group is called *oligomorphic*. (The term is intended to suggest "few shapes", referring to a structure on Ω admitting G as a group of automorphisms — there are only finitely many, i.e. "only a few", different shapes of n-element subsets.)

(The orbits of G on Ω^n refer to the obvious induced action of G, defined by

$$(\alpha_1, \ . \ . \ , \alpha_n)g = (\alpha_1 g, \ldots, \alpha_n g)$$

for $\alpha_1, \ldots, \alpha_n \in \Omega, g \in G$. Similarly, there are obvious actions of G on various other sets related to Ω that we consider below, such as subsets of fixed size, partitions, graphs with vertex set Ω, etc. The action will not be spelt out each time.)

An equivalent requirement is that G has only finitely many orbits on the set of n-tuples of distinct elements of Ω, or on the set of n-element subsets of Ω, for each n. Before continuing with the serious business, I shall briefly consider these numbers of orbits and their relationships.

Let f_n, F_n, F_n^* denote the numbers of orbits of G on n-subsets, n-tuples of distinct elements, and arbitrary n-tuples, respectively. We will write $f_n(G)$, etc., if it is necessary to specify the group G. (To be consistent, Ω should be included in the notation as well; I will not do this.) By convention, $f_0 = F_0 = F_0^* = 1$.

There isn't a precise relationship between f_n and F_n, but we have the following result:

(2.1) $f_n \leq F_n \leq n!f_n,$

because a given n-set has $n!$ different orderings which lie in somewhere between 1 and $n!$ different orbits.

We say that G is *highly homogeneous* if $f_n = 1$ for all n; it is *highly transitive* if $F_n = 1$ for all n. Two simple but important examples illustrate these concepts:

(i) The symmetric group $\mathrm{Sym}(\Omega)$ is highly transitive (and, *a fortiori*, highly homogeneous).

(ii) The group $\mathrm{Aut}(\mathbf{Q}, <)$ of order-preserving permutations of $\mathbf{Q}$ is highly homogeneous. For, given any two n-tuples $(a_1, \ldots, a_n)$ and $(b_1, \ldots, b_n)$, both in strictly increasing order, we can map the first to the second, interpolate a linear function in each interval (a_i, a_{i+1}), and use a shift on each "end", to define an element of $\mathrm{Aut}(\mathbf{Q}, <)$. This group is not highly transitive, since no order-preserving permutation may interchange two distinct points. Indeed, $f_n = 1$ and $F_n = n!$ for all n.

These two examples show that both bounds in (2.1) can be attained. But more information can be deduced if the first bound is met:

(2.2) *If $f_{n+1} = F_{n+1}$, then $F_n = 1$.*

For the hypothesis implies that any permutation of $n + 1$ points can be induced by an element of G. So, if two n-tuples of distinct elements share $n - 1$ points, then we can map one to the other by an element of G which fixes their union setwise. Then we can map any n-tuple to any other by a sequence of steps of this kind, changing one point at a time.

A permutation group is called n-*homogeneous* if $f_n = 1$, n-*transitive* if $F_n = 1$, and *generously n-transitive* if $F_{n+1} = f_{n+1}$. (The term "homogeneous" is unfortunate here, since it conflicts with a more important (in this context) use of the same word, which will be introduced in §2.5.)

The relationship between F_n and F_n^* is much more precise. It involves the *Stirling numbers of the second kind*, a celebrated array of numbers defined by the rule
$$S(n, k) = \text{the number of partitions of an } n\text{-set into } k \text{ parts.}$$

(Note that we are partitioning a set, not an integer, here.)

(2.3) $F_n^* = \sum_{k=1}^n S(n,k)F_k$.

To see this, note that an n-tuple $(\alpha_1, \ldots, \alpha_n)$ defines a partition of $\{1, \ldots, n\}$, in which i and j lie in the same part if and only if $\alpha_i = \alpha_j$. Two n-tuples lie in the same orbit of G if and only if
   (a) the partitions they define are equal;
   (b) the tuples formed by their distinct elements (in order of first appearance) lie in the same orbit of G.

This relationship can also be expressed in terms of the *exponential generating functions* (e.g.f.s) of the sequences, the formal power series

$$F(t) = \sum_{n \geq 0} F_n t^n / n!$$

and

$$F^*(t) = \sum_{n \geq 0} F_n^* t^n / n!$$

(2.4) $F^*(t) = F(e^t - 1)$.

For more details, applications and generalisations, see Cameron and Taylor (1985), from which the following are taken.

Exercises

1. The *Bell number $B(n)$* is the number of partitions of an n-set. Prove that, for any permutation group G, we have $F_n^* \geq B(n)$, with equality if and only if G is n-transitive. Show also that the e.g.f. for $B(n)$ is $e^{e^t - 1}$.

2. A *preorder* is a reflexive and transitive relation. By applying (2.3) and (2.4) to the group $\mathrm{Aut}(\mathbf{Q}, <)$, show that the number of preorders of an n-set is

$$\sum_{k=1}^n S(n,k)k!,$$

with e.g.f. $1/(2 - e^t)$.

3. Prove that, for any oligomorphic permutation group G,

$$F_n^*(G) = F_n(S \,\mathrm{Wr}\, G),$$

where S is the symmetric group on a countably infinite set, and the wreath product has its imprimitive action (see §1.2).

2.2. REDUCTION TO THE COUNTABLE CASE

The question which originally motivated my interest in the subject was this:

Given a sequence of positive integers, can it be realised as f_n (or F_n) for some oligomorphic permutation group?

In order to study this question, we need not go beyond permutation groups of countable degree. There is no need to grapple with the uncertainties of large cardinal numbers! This is a consequence of the downward Löwenheim-Skolem theorem (1.6). This theorem asserts that, if a set of sentences in a countable first-order language has a model, then it has a countable model. So the claim will be justified by noting that all relevant information about the group can be cast into sentences of first-order logic.

To do this, we let the domain consist of two kinds of elements, the points of Ω and the elements of G; we include in the language a unary predicate to distinguish them. We also include the usual language of group theory (i.e. functions for multiplication and inversion, and a constant for the identity), and a binary function for the action of G on Ω. The theory includes the axioms for a group, and those for a G-space. For any natural number k, there is a sentence asserting that G has at least k orbits on Ω ("there exist k points such that no element of G maps any one to any other"), and hence a sentence asserting that there are exactly k orbits. Simple modifications say the same thing for n-sets, or for n-tuples. So all the data can be expressed.

More recondite information can also be expressed; for example, the modified cycle index function to be defined in §3.7, or more generally the list of permutation groups induced on finite subsets by their setwise stabilisers.

Primitivity cannot be expressed by first-order sentences in general (see Exercise 1). But, if we know F_2, then we can say that G is primitive in the following way. In general, a transitive group G is primitive if and only if, for any orbit O of G on ordered pairs of distinct elements, the symmetric and transitive closure of O is the universal relation. In other words, given $\alpha, \beta, \gamma, \delta$ with $\gamma \neq \delta$, there is a sequence $(\gamma_0, \ldots, \gamma_n)$ with $\gamma_0 = \alpha$, $\gamma_n = \beta$, and for $i = 0, \ldots, n-1$, either (γ_i, γ_{i+1}) or (γ_{i+1}, γ_i) lies in the G-orbit of (γ, δ). (See Higman (1967).) Now, if F_2 is known, then the existence of such a sequence is equivalent the existence of one with $n \leq F_2$. For, given γ and δ, we can define the *distance* from α to β to be the least n for which such a sequence exists. Then the realised distances form an initial segment of the natural numbers; and two pairs of points lying at different distances are certainly in different orbits, so no more than $F_2^* = F_2 + 1$ different distances can occur (including zero).

From the Löwenheim-Skolem theorem, we conclude that we may assume that both Ω and G are countable. The countability of Ω is more significant than that of G. There are two reasons for this.

Firstly, given any permutation group G on a countable set Ω, there is a countable subgroup with the same orbits on tuples as G, for reasons much more elementary than the Löwenheim-Skolem theorem. We can enumerate the pairs $(\bar{a}_n, \bar{b}_n)$ of tuples for which $\bar{a}_n$ and $\bar{b}_n$ lie in the same G-orbit, and choose an element g_n of G carrying $\bar{a}_n$ to $\bar{b}_n$ for each n; the subgroup generated by the chosen elements is countable and has the required property.

Secondly, while it is to our advantage to require countability of Ω, there are good reasons (related to topological considerations) for not requiring this of G. Note, for example, that the symmetric group $\mathrm{Sym}(\Omega)$ has cardinality $2^{\aleph_0}$, as do the automorphism groups of many interesting countable structures.

From now on, I always assume that Ω is countable.

Exercises

1. Prove that there is no set Σ of sentences (in the language of permutation groups described above) with the property that Σ characterises primitive groups. (*Hint:* Regularity is first-order; but primitivity and regularity together imply finiteness. See Exercise 2 of §1.2 and the remarks preceding (1.7).)

A harder exercise along the same lines is the following: Construct pairs (G_1, Ω_1) and (G_2, Ω_2) which satisfy the same first-order sentences, such that G_1 acts primitively on Ω_1 and G_2 imprimitively on Ω_2.

2. Show that the distance defined in this section is a metric on Ω.

3. Let G be transitive on Ω. Show that there is a bijection from the set of G-orbits on Ω^2 to the set of G_α-orbits on Ω, defined by

$$O \mapsto O(\alpha) = \{\beta \in \Omega : (\alpha, \beta) \in O\}.$$

Show that $O(\alpha)g = O(\alpha g)$ for all $g \in G$.

Hence show that

(i) if G is primitive on Ω, and G_α has a finite or countable orbit (other than $\{\alpha\}$, then Ω is finite or countable;

(ii) if G is primitive on Ω, $F_2(G)$ is finite, and G_α has a finite orbit Δ (other than $\{\alpha\}$), then Ω is finite.

In (ii), find an upper bound for $|\Omega|$ in terms of $d = F_2(G)$ and $k = |\Delta|$.

4. Prove that, if the value of $F_2(G)$ is given, then the statement "G is contained in the automorphism group of a linear order on Ω" is first-order.

More generally, if $F_n(G)$ is given, then a first-order sentence asserts that G acts on a relational structure with an n-ary relation satisfying some first-order axioms.

Is this true without the assumption that F_n is given?

2.3. THE CANONICAL RELATIONAL STRUCTURE

(2.5) *If G is any permutation group on a countable set Ω, then there is a relational structure M on Ω (over a countable relational language) such that*
(i) $G \leq \mathrm{Aut}(M)$;
(ii) G and $\mathrm{Aut}(M)$ have the same orbits on Ω^n, for every n.

The structure which I construct to demonstrate this is called the *canonical relational structure* associated with G. Decompose $\bigcup_{n \geq 1} \Omega^n$ into orbits $O_1, O_2, \ldots$; with each orbit O_i associate a relation symbol R_i, of arity n if $O_i \subseteq \Omega^n$; and interpret R_i as O_i. Then (i) is clear. Any tuple satisfies R_i for a unique value of i; so, if $\bar{a}$ and $\bar{b}$ lie in the same $\mathrm{Aut}(M)$-orbit, then they satisfy the same relation R_i, and so lie in the same G-orbit O_i.

Of course, G may be the automorphism group of a structure over quite a different language, perhaps one with only finitely many relation symbols (as in our example $\mathrm{Aut}(\mathbf{Q}, <)$), or one with function and constant symbols too. Moreover, a variant of the canonical structure, which is just as good for our purposes and more economical with relation symbols, is obtained by considering only orbits on tuples with distinct elements.

Exercise

Let G be a permutation group on a countable set Ω. Suppose that, for some n, every non-identity element of G fixes fewer than n points. Prove that G is the automorphism group of a relational structure over a language with a single relation, whose arity can be taken to be $2n + 2$.

[*Hint.* Enumerate the orbits of G on $n + 1$-tuples; take the relation which asserts that its first $n + 1$ arguments form a tuple in the ith orbit and its last $n + 1$ a tuple in the $(i + 1)^{\text{st}}$, for some i.]

2.4. TOPOLOGY

There is a natural topology defined on the symmetric group, that of *pointwise convergence*. (Identify Ω with $\mathbf{N}$, the natural numbers; in other words, enumerate $\Omega = (\alpha_0, \alpha_1, \ldots)$. Then a sequence (g_n) of permutations tends to the limit g if and only if, for any $k \in \mathbf{N}$, $\alpha_k g_n = \alpha_k g$ for all sufficiently large n.)

This makes G into a *topological group*. (This is just to say that multiplication and inversion are continuous; explicitly, if $g_n \to g$ and $h_n \to h$, then $g_n h_n \to gh$ and $g_n^{-1} \to g^{-1}$.)

A basis for the open sets in this topology consists of the cosets of stabilisers of tuples: these are the sets
$$\{g \in \text{Sym}(\Omega) : \bar{a}g = \bar{b}\}$$
where $\bar{a}$ and $\bar{b}$ are tuples of distinct elements of the same length. (This set is empty unless these tuples lie in the same orbit, of course.)

The topology can be derived from a metric. Suppose that $\Omega = \mathbf{N}$. Set
$$d(g, h) = \begin{cases} 0, & \text{if } g = h; \\ 1/2^i, & \text{if } jg = jh \text{ for all } j < i \text{ but } ig \neq ih. \end{cases}$$

This is very natural; the larger initial segment of $\mathbf{N}$ on which g and h agree, the closer they are. In view of the continuity of inversion in this topology, which the reader should stop and prove if (s)he has not already done so, the same topology is obtained from the metric d' defined by
$$d'(g, h) = \begin{cases} 0, & \text{if } g = h; \\ 1/2^i, & \text{if } jg = jh \text{ and } jg^{-1} = jh^{-1} \text{ for all } j < i, \text{ but} \\ & ig \neq ih \text{ or } ig^{-1} \neq ih^{-1}. \end{cases}$$

In other words,
$$d'(g, h) = \max(d(g, h), d(g^{-1}, h^{-1})).$$
The advantage of this more complicated-seeming metric is that it makes $\mathrm{Sym}(\Omega)$ into a complete metric space; this is false for the metric d.

(2.6) *A subgroup G of $\mathrm{Sym}(\Omega)$ is closed if and only if $G = \mathrm{Aut}(M)$ for some (first-order) structure M on Ω.*

For suppose that $g_n \in \mathrm{Aut}(M)$ ($n \in \mathbf{N}$) and $g_n \to g$. For any tuple $\bar{a}$, there exists n such that $\bar{a}g = \bar{a}g_n$; thus $\bar{a}g$ satisfies a relation of M if and only if $\bar{a}$ does (since g_n is an automorphism of M). It follows that $g \in \mathrm{Aut}(M)$; so $\mathrm{Aut}(M)$ is closed. In the other direction, suppose that G is closed, and let M be its canonical structure and $g \in \mathrm{Aut}(M)$. For any tuple $\bar{a}$, there exists $g' \in G$ such that $\bar{a}g = \bar{a}g'$ (by (2.5)(ii)). Let g_n be the element g' obtained thus when $\bar{a} = (0, \ldots, n-1)$. Then $g_n \to g$. Since G is closed, $g \in G$.

The argument shows, in fact, that

for any permutation group G, the closure of G in $\mathrm{Sym}(\Omega)$ is the automorphism group of the canonical structure of G.

In particular, if G is the automorphism group of any first-order structure, then it is the automorphism group of its canonical structure.

This result is not valid for other kinds of mathematical structure on Ω. For example, the group of topological homeomorphisms of the topological space $\mathbf{Q}$ is highly transitive, and so its closure is the symmetric group. (See Exercise 3. Neumann (1985a) has made a detailed study of this permutation group.)

(2.7) *For any countably infinite first-order structure M, either $|\mathrm{Aut}(M)| \leq \aleph_0$ or $|\mathrm{Aut}(M)| = 2^{\aleph_0}$, the first alternative holding if and only if the stabiliser of some tuple is the identity.*

If the stabiliser of some tuple in $G = \mathrm{Aut}(M)$ is the identity, then G is equal to the number of tuples in its orbit, and so is at most countable. (In this case, the induced topology on G is discrete.) Otherwise, the identity, and hence every point, is a limit point, and $|G| = 2^{\aleph_0}$.

A refinement of (2.7), due to Evans (1987b), is closely related to Kueker's definability theorem (1968):

(2.8) *If G and H are closed subgroups of $\mathrm{Sym}(\Omega)$, and $G \leq H$, then either $|H : G| \leq \aleph_0$ or $|H : G| = 2^{\aleph_0}$. The former holds if and only if G contains the stabiliser (in H) of some tuple.*

I will show only the second part This uses an important tool, the Baire Category Theorem, which applies since H (as a closed subspace of a complete metric space) is itself complete. Assume that $|G : H| \leq \aleph_0$. Then H is the union of countable many cosets of G; by the Baire Category Theorem, some coset has non-empty interior. By translation, G has non-empty interior, and every element of G (in particular, the identity) is in the interior. A basic open set containing the identity is the stabiliser of a tuple.

A pleasant corollary of (2.8) is

(2.9) *Let H be an oligomorphic closed subgroup of $\mathrm{Sym}(\Omega)$. If G is a subgroup of H of index less than $2^{\aleph_0}$, then G is also oligomorphic.*

For the property in question is preserved by (among other things) descending to the stabiliser of a tuple, and passing to a supergroup; and we can replace G by its closure without increasing its index.

The presentation in this section has been lifted (with thanks) from David Evans.

Exercises

1. Show that the sequence (g_n) where

$$g_n = (0\ 1\ 2\ \ldots\ n - 1),$$

is a Cauchy sequence for the metric d which fails to converge. (Its pointwise limit is not a permutation.) If this example makes you a bit nervous, go back and verify the assertions in the text about the topology of $\mathrm{Sym}(\Omega)$.

2. Let H be oligomorphic and closed in $\mathrm{Sym}(\Omega)$, and G a subgroup of countable index in H. Show that there exists an integer d such that $F_n(G) \leq F_{n+d}(H)$ for all n.

3. Show that the map

$$x \mapsto \begin{cases} x, & \text{if } x^2 > 2; \\ -x, & \text{otherwise} \end{cases}$$

is a homeomorphism of $\mathbf{Q}$. Hence show that the group of homeomorphisms is highly transitive.

Harder: Show that $\mathbf{Q}$ is homeomorphic to $\mathbf{Q}^2$, and hence show that Homeo($\mathbf{Q}$) is highly transitive.

See Neumann (1985a).

2.5. THE RYLL-NARDZEWSKI THEOREM

We saw in §1.3 that no infinite structure can be described completely by first-order axioms. (The upward Löwenheim-Skolem theorem (1.7) asserts that a theory with an infinite model has arbitrarily large infinite models.) As we are only interested in countable structures, a reasonable compromise is to admit a single non-first-order assumption, that of countability. A complete theory is $\aleph_0$-*categorical* if it has, up to isomorphism, a unique countable model. Moreover, we call a countable structure $\aleph_0$-*categorical* if its theory is. The prototype is Cantor's theorem:

Any countable dense linearly ordered set without endpoints is order-isomorphic to $\mathbf{Q}$.

(The axioms for linear order, denseness, and absence of end-points are first-order.)

It is not at all obvious that $\aleph_0$-categoricity of a structure is "really" a property of its automorphism group, and indeed, a property implying a very high degree of symmetry! (Perhaps consideration of Felix Klein's Erlanger Programm might lead us to suspect a connection between axiomatisability and symmetry.) This remarkable discovery was made independently in 1959 by Engeler, Ryll-Nardzewski and Svenonius. Though it usually has the second of these names attached to it, the form given by Svenonius is closest to my point of view here.

   (2.10) *A countable first-order structure M is* $\aleph_0$-*categorical if and only if* Aut(M) *is oligomorphic.*

The proof of the backward implication (and somewhat more) will be given later, in §2.8. Here, I want to draw attention to one feature of the proof. The *type* of a tuple $\bar{a}$ in a structure M (written $\text{tp}_M(\bar{a})$) is the collection of all first-order formulae $\phi(\bar{x})$ (in a tuple $\bar{x}$ of free variables of the same arity as $\bar{a}$) such that $\phi(\bar{a})$ holds in the structure M. (*Note:* This is not the general definition of a type in first-order logic. It will emerge in the proof of (2.10) that, in the special case of an $\aleph_0$-categorical theory, this definition agrees with the more general form.)

Now, in the course of the proof, we show the following. Let M and N be models of an $\aleph_0$-categorical theory T, and $\bar{a}$ and $\bar{b}$ tuples in M and N respectively. Then:

($\Diamond$) *If* $\mathrm{tp}_M(\bar{a}) = \mathrm{tp}_N(\bar{b})$ *and* $x \in M$, *then there exists* $y \in N$ *such that* $\mathrm{tp}_M(\bar{a}, x) = \mathrm{tp}_N(\bar{b}, y)$.

(Here $(\bar{a}, x)$ denotes the tuple consisting of the terms of $\bar{a}$ followed by x.)

This is exactly what is required to operate a machine familiar to model theorists: the *back-and-forth argument*. Enumerate the domains of M and N, as $(x_0, x_1, \ldots)$ and $(y_0, y_1, \ldots)$ respectively, and define a map $\theta : M \to N$ thus. At even-numbered stages, let x_i be the first x (in the enumeration) which is not in the domain $\bar{a}$ of θ (as defined so far); choose $y \in N$ such that $\mathrm{tp}_M(\bar{a}, x_i) = \mathrm{tp}_N(\bar{a}\theta, y)$. (To avoid any suspicion of having invoked a choice principle, let y be the first y_j in the enumeration of N to satisfy this condition.) Extend θ by mapping x_i to y. At odd-numbered stages, take the first y_j not in the range of θ, and choose a pre-image in M. The point of going back-and-forth in this way is to ensure that the domain and range of the map θ constructed after infinitely many stages are all of M and N — the nth elements are included no later than stage $2n + 1$. Since the type of a tuple includes complete information about its isomorphism type, θ is "locally" an isomorphism, and hence is an honest isomorphism. So $M \cong N$. But the argument gives even more. We can apply it with M and N replaced by $(M, \bar{a})$ and $(M, \bar{b})$ whenever $\mathrm{tp}_M(\bar{a}) = \mathrm{tp}_M(\bar{b})$. (Here $(M, \bar{a})$ means the structure consisting of M with the terms of $\bar{a}$ as distinguished constants, by means of new constant symbols added to the language of M.) We conclude that there is an automorphism of M carrying $\bar{a}$ to $\bar{b}$. Thus, the orbits of $\mathrm{Aut}(M)$ on tuples are precisely the types, and there are only finitely many such orbits on n-tuples for all n.

To reiterate: Whenever condition ($\Diamond$) holds, then the structures M and N are isomoorphic, and two tuples in M lie in the same orbit of the automorphism group of M if and only if they have the same type. Furthermore, the precise definition of "type" is not too important here. Any equivalence relation on tuples satisfying ($\Diamond$) would do, provided that we interpret "isomorphism" to mean "type-preserving bijection".

We can re-phrase the Ryll-Nardzewski theorem to refer to arbitrary permutation groups:

The group G is oligomorphic if and only if its canonical structure is $\aleph_0$-categorical.

Note. The back-and-forth argument will be used again in the next section. Later, in §5.2, I will make a few remarks on its history, and discuss a question of Adrian

Mathias:

Is back-and-forth really necessary, or would forth alone suffice?

Exercise

It follows from the Ryll-Nardzewski theorem that $(\mathbf{Z}, <)$ is not $\aleph_0$-categorical. Describe the countable models of its theory. How many are there?

2.6. HOMOGENEOUS STRUCTURES

In this section, it is convenient to work with a relational language. For structures over relational languages have the property that any subset of the domain carries a substructure; this is important if we wish to describe all orbits on finite subsets. There are, however, more general versions of the main theorem below.

First, some definitions.

(i) The *age* of a structure M, written Age(M), is the class of all structures (over the same language as M) which are isomorphic to finite substructures of M. (Of course, Age(M) is a proper class, not a set. This doesn't usually matter; and in any case, it is often resolved by the fact that we only need to consider members of Age(M) whose domain consists of a subset of the natural numbers, and these do form a set. If you don't understand this comment, there is no need to worry.)

(ii) A class $\mathcal{C}$ of finite structures has the *amalgamation property* if, whenever A, B_1, $B_2 \in \mathcal{C}$ and $f_i : A \to B_i$ are embeddings ($i = 1, 2$), then there exist $C \in \mathcal{C}$ and embeddings $g_i : B_i \to C$ ($i = 1, 2$) so that the diagram

$$
\begin{array}{ccc}
A & \xrightarrow{f_1} & B_1 \\
{\scriptstyle f_2}\downarrow & & \downarrow{\scriptstyle g_1} \\
B_2 & \xrightarrow{g_2} & C
\end{array}
$$

commutes. (Informally, if B_1 and B_2 have isomorphic substructures, they can be glued together along these substructures, possibly with some extra identifications.)

(iii) A structure M is *homogeneous* if any isomorphism between finite substructures of M can be extended to an automorphism of M. For example, $\mathbf{Q}$ (as ordered set) is homogeneous: this is shown by the argument in §2.1, where we observed that

Aut($\mathbf{Q}$, $<$) is highly homogeneous (and I warned about conflicting terminology). This example is the prototype for the theory below. Also,

(2.11) *The canonical structure for any permutation group is homogeneous.*

For, if $\bar{a}$ and $\bar{b}$ are the domain and range of a finite isomorphism, then the unique relation R_i satisfied by $\bar{a}$ also holds for $\bar{b}$, so $\bar{a}$ and $\bar{b}$ lie in the same G-orbit.

If M is homogeneous, then $f_n(\mathrm{Aut}(M))$ is the number of isomorphism types of n-element structures in the age of M, while $F_n(\mathrm{Aut}(M))$ is the number of members of the age with a fixed domain of cardinality n. (In the terminology of combinatorial enumeration, these are the numbers of *unlabelled* and *labelled* structures, respectively, in Age(M). This observation is the link between our concerns here and combinatorial enumeration theory, for which see Goulden and Jackson (1983) or Harary and Palmer (1973), for example.)

The tool for constructing homogeneous structures is Fraïssé's theorem. I have separated the existence and uniqueness statements of the theorem.

(2.12) *The class C of finite structures over a relational language is the age of some countable homogeneous relational structure if and only if it is closed under isomorphism and under taking substructures, has only countably many non-isomorphic members, and has the amalgamation property.*

(Incidentally, the same result holds with "finite" and "finitely" replacing "countable" and "countably".)

(2.13) *Two countable homogeneous structures with the same age are isomorphic.*

All the conditions of (2.12) are necessary; but in practice, the first three are mere bookkeeping, and only the amalgamation property requires any real work in its verification. We could enforce the third condition by requiring that the language contains only finitely many "essential" relations of any given arity; but, as well as being difficult to formulate precisely, this would be a real loss of generality: it would exclude a construction technique which we will exploit later.

Again, the proof is deferred to §2.8, but I draw attention to one feature. The uniqueness and the homogeneity of the structure we construct in the proof both follow by back-and-forth from the following assertion, where M and N are structures and $\bar{a}$ and $\bar{b}$ are tuples in M and N respectively:

($\heartsuit$) *If $\bar{a}$ and $\bar{b}$ carry isomorphic substructures, and $x \in M$, then there exists $y \in N$ such that $(\bar{a}, x)$ and $(\bar{b}, y)$ carry isomorphic substructures.*

(This is the same as ($\Diamond$) but with isomorphism type replacing first-order type — but, as we remarked, the exact definition of type is not important.)

The uniqueness statement (2.13) is not true if the condition of homogeneity is replaced by $\aleph_0$-categoricity. Droste and Macpherson (to appear) have constructed $2^{\aleph_0}$ non-isomorphic countable $\aleph_0$-categorical graphs, the age of each graph being the class of all finite graphs.

Now for some examples. The first one resolves the question: which degrees of transitivity are realised by infinite permutation groups? (This question is motivated by the result, which was an empirical observation for a century until it was established when the classification of the finite simple groups was completed in 1980: apart from the symmetric and alternating groups, there is no 6-transitive finite permutation group.)

(2.14) *For any k, there is a permutation group which is k-transitive but not $(k+1)$-transitive.*

A $(k+1)$-*uniform hypergraph* consists of a set of vertices and a set of "hyperedges", each hyperedge a $(k+1)$-subset of the vertex set. (Ordinary graphs are obtained when $k = 1$.) The class $\mathcal{R}_k$ of all finite $(k+1)$-uniform hypergraphs satisfies Fraïssé's conditions. (To form the amalgam of two hypergraphs, make no unnecessary identifications and add no new hyperedges.) Let R_k be the homogeneous object guaranteed by (2.12). Since a $(k+1)$-uniform hypergraph with k vertices is trivial, any bijection between k-sets extends to an automorphism; so $G = \mathrm{Aut}(R_k)$ is k-transitive. But it is not $(k+1)$-transitive, since no automorphism can map edges to non-edges.

A less trivial example is the following. A *boron tree* is a finite tree in which all vertices have valency 1 ("hydrogen atoms") or 3 ("boron atoms"). (These are the analogues of hydrocarbons in an imaginary chemistry in which trivalent boron replaces tetravalent carbon.) Let $\mathcal{B}$ be the class of structures (over a language with one quaternary relation R) defined as follows. The points are the H atoms of some finite boron tree; $R(a, b, c, d)$ holds when the paths joining a to b and c to d are disjoint. (This holds for exactly one of the three partitions of the 4-set into two 2-sets; see Fig. 1.) It can be shown that $\mathcal{B}$ has the amalgamation property; the other parts of Fraïssé's hypotheses are (as usual) trivial. If B is the countable homogeneous object, then $G = \mathrm{Aut}(B)$ is 5-homogeneous and 3- but not 4-transitive, and satisfies $f_6 = f_7 = 2$. These assertions can all be read off from the list of isomers of boron trees with at most seven H atoms. Those with at most six are shown in Fig. 2.

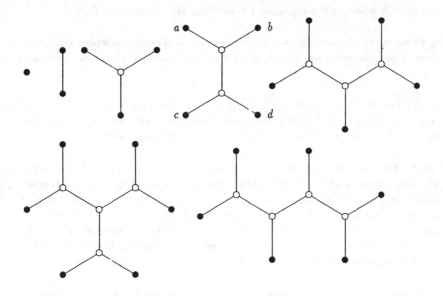

Fig. 2. Boron trees

Several generalisations of this example are found in Cameron (1987b).

Exercises

1. Prove that the class C of finite structures is the age of some countable structure if and only if C is closed under isomorphism and under taking substructures, has only countably many non-isomorphic members, and has the *joint embedding property*, viz. given any $A, B \in C$, there exists $C \in C$ into which both A and B can be embedded.

Continuation. Show that the countable structure is not unique up to isomorphism. Show directly that the joint embedding property is a consequence of the amalgamation property, but that this would be false if the empty set was not included as a structure. Finally, formulate and prove a finite version of this result.

2. For every positive integer $k > 1$, construct a permutation group which is k-homogeneous but not k-transitive.

3. For every positive integer $k > 1$, construct a k-transitive group having a subgroup

of index 2 which is not k-transitive. [*Hint:* R_k is isomorphic to its *complement*, the hypergraph whose edges are those k-sets which are not edges of R_k.]

4. Prove the amalgamation property for the class $\mathcal{B}$ of quaternary relational structures derived from boron trees. [*Hint:* Show that the relational structure determines the tree.]

5. (i) Let M be a countable $\aleph_0$-categorical L-structure, where L is a relational language. Suppose that N is a countable L-structure with $\mathrm{Age}(N) \subseteq \mathrm{Age}(M)$. (We say that N is *younger than* M.) Show that N is embeddable in M.

[*Hint.* Let the point set of N be the set of natural numbers. Construct a tree whose nodes at level n are equivalence classes of embeddings $N_n \to M$, where N_n is the induced substructure of N on $n = \{0, \ldots, n-1\}$. (Two embeddings f_1, f_2 are *equivalent* if $f_2 = f_1 g$ for some $g \in \mathrm{Aut}(M)$.) Adjacency between nodes on consecutive levels is defined by restriction. By the Ryll-Nardzewski theorem, there are only finitely many nodes at level n. Then König's Infinity Lemma shows that the tree has an infinite path. Use this path to define an embedding of N into M.]

(ii) Same question with "homogeneous" replacing "$\aleph_0$- categorical". (This is much easier.)

(iii) It is not always true that every structure younger than M is embeddable in M. Give an example.

2.7. STRONG AMALGAMATION

I will on several occasions need a stronger version of the amalgamation property. The *strong amalgamation property* is the assertion that, in the notation of the amalgamation property (§2.6), C, g_1 and g_2 can be chosen so that, if $b_i \in B_i$ ($i = 1, 2$) with $b_1 g_1 = b_2 g_2$, then there exists $a \in A$ with $b_i = a f_i$ ($i = 1, 2$). Said more simply, the amalgam can be formed without making any identifications other than those specified.

Let G be a permutation group on Ω. The *algebraic closure* $\mathrm{acl}(A)$ of a finite subset A of Ω is the set of all those points of Ω which lie in finite orbits of the pointwise stabiliser of A. So, for example, if Ω carries the structure of a projective or affine space over a finite field, and G is its collineation group, then the algebraic closure of a finite set is the subspace it spans. When I speak of algebraic closure in a first-order structure, I take the group in question to be the automorphism group of the structure. (*Note:* there is a model-theoretic definition of algebraic closure. For $\aleph_0$-categorical structures, it coincides with the definition I've given here, essentially because first-order types coincide with orbits of the automorphism group in this case.)

(**2.15**) *Let M be a homogeneous structure. Then the following are equivalent:*
(i) the age of M has the strong amalgamation property;
(ii) for any finite set $A \subseteq M$, $\operatorname{acl}(A) = A$.

This result, or something like it, has two lines of descent, in permutation group theory and in model theory. On the group-theoretic side, its original ancestor is a theorem of B. H. Neumann (1954), according to which, if a group can be expressed as the union of finitely many cosets of (possibly distinct) subgroups, then one at least of these subgroups has finite index. From this, Π. M. Neumann (1976) deduced:

(**2.16**) *Suppose that G has no finite orbits on Ω, and let A and B be finite subsets of Ω. Then there exists $g \in G$ with $Ag \cap B = \emptyset$.*

Later, with three others (Birch, Burns, Macdonald and Neumann(1976)), he published a direct proof. Here is a slightly different proof. The argument is a nested induction. The outer induction is on $|A|$; so we assume the result for any set A' with $|A'| < |A|$. (The induction trivially begins when $A = \emptyset$.) For a contradiction, we assume that no element g with the required property exists.

Claim: For any set C with $|C| \leq |A|$, only finitely many translates of A contain C.

The claim is proved by induction on $|A| - |C|$; it starts trivially when $|A| - |C| = 0$, so suppose that $|C| < |A|$ and that the claim holds for all C' with $|C'| > |C|$. Then, by the outer induction hypothesis, we may assume (translating C if necessary) that $C \cap B = \emptyset$. By the inner induction hypothesis, for each of the (finitely many) points $b \in B$, only finitely many translates of A contain $C \cup \{b\}$. So only finitely many translates of A contain C and meet B. But, since we have a counterexample to (2.16), every translate of A meets B. So the inner induction goes through, and the claim is proved.

Now, taking $C = \emptyset$ in the claim, we see that A has only finitely many translates. But this is a clear contradiction to the hypothesis that G has no finite orbits. So the result is proved.

This result is used to prove that (ii) implies (i) in (2.15). For assume (ii), and let A, B_i, f_i ($i = 1, 2$) be as in the hypotheses of the amalgamation property. By homogeneity, we can assume that $A \subseteq B_1 \cap B_2 \subseteq M$, and that f_1 and f_2 are the identity map on A. Now $G_{(A)}$ (the pointwise stabiliser of A) has no finite orbits outside A. By (2.16), there exists $g \in G_{(A)}$ such that $(B_1 \setminus A)g \cap (B_2 \setminus A) = \emptyset$. Then $B_1 g \cup B_2$ is the required strong amalgam.

To show that (i) implies (ii), take any finite set $A \subseteq M$, and use strong amalgamation to show that, for any natural number n and any given point $x \in M \setminus A$, there exist n distinct points having the same type as x over A. By homogeneity, they all lie in the same $G_{(A)}$-orbit as x. So the orbit of x is infinite. Since x was arbitrary, $G_{(A)}$ has no finite orbits outside A.

Exercises

1. Show that (2.16) is equivalent to B. H. Neumann's result cited in the text (in the sense that each can be deduced from the other more easily than it can be proved directly).

2. For any m, n, if all G-orbits have size greater than mn, and $|A| = m$, $|B| = n$, then there exists $g \in G$ such that $Ag \cap B = \emptyset$. Show that this is best possible. *Harder:* Prove the assertion. (See Birch *et al.* (1976)).

3. The following theorem was proved by Wielandt (1959), and is an infinite version of a theorem of Jordan about finite permutation groups:

A primitive infinite permutation group which contains a non-identity element of finite support contains the alternating group.

(Here, the *support* supp(g) of a permutation g is the set of points it moves; the *alternating group* consists of all permutations with finite support which act on their supports as even permutations) The proof outlined below was shown me by John Dixon.

Step 1. If g moves α, then g moves a point in every orbit of G_α. For, if g fixed $O(\alpha)$ pointwise, then the relation defined by $\gamma \sim \delta$ if $O(\gamma) = O(\delta)$ (cf. Exercise 3 of §2.2) is a congruence, and is non-trivial (since $\alpha \sim \beta$). Hence G_α has only finitely many orbits.

Step 2. By Exercise 3(ii) of §2.2, all the orbits of G_α except $\{\alpha\}$ are infinite.

Step 3. Now by (2.16), there exists $h \in G_\alpha$ such that supp(g) $\cap$ supp(g)$h = \{\alpha\}$. This implies that $g^{-1}h^{-1}gh$ is a 3-cycle.

Step 4. Finally, G contains all 3-cycles, and hence contains the alternating group.

[The theorem of Jordan asserts the existence of a function f on $\mathbf{N}$, tending to infinity with its argument, such that a finite primitive permutation group of degree n and containing a non-identity element whose support has size at most $f(n)$, contains the alternating group.]

2.8. APPENDIX: TWO PROOFS

In this section, I give, in a little more detail, the proofs of the theorems of Ryll-Nardzewski (in one direction) and Fraïssé.

The Ryll-Nardzewski theorem (2.10) follows from the next two results. First, the promised general definition of types. Let Σ be a complete consistent theory (that is, the set of all sentences true in some structure). An *n-type* over Σ is a set of formulae $\phi(\bar{x})$ (in an n-tuple $\bar{x}$ of free variables) maximal consistent subject to the sentences $(\exists\bar{x})\phi(\bar{x})$ being consistent with Σ. (The type of an n-tuple defined in §2.5 is a special case. We say that a type is *realised* in a model M of Σ if it is the type of some n-tuple in M. Note that two n-tuples which realise different types necessarily lie in different orbits of the automorphism group of M. It is a consequence of Gödel's completeness theorem (1.4) that every type is realised in some model of Σ.)

(2.17) *Let Σ be a complete consistent theory having only finitely many n-types for all n. Then*
(i) every type is realised in any model of Σ;
(ii) any two countable models of Σ are isomorphic;
(iii) if M is a countable model of Σ, and $\bar{a}$ and $\bar{b}$ are n-tuples in M having the same type, then there is an automorphism of M carrying $\bar{a}$ to $\bar{b}$.

(2.18) *Let Σ be a complete consistent theory having infinitely many n-types for some n. Then Σ has two non-isomorphic countable models.*

(In fact, Vaught (1963) showed that, under the hypotheses of (2.18), Σ has at least three non-isomorphic countable models, and this is best possible — every natural number except 2 is realised!)

I will prove the first of these two results.

Proof of (2.17). For a fixed n, let $T_1, \ldots, T_r$ be the n-types. For $i \neq j$, there is a formula $\phi_{ij}(\bar{x})$ lying in either $T_i \setminus T_j$ or $T_j \setminus T_i$; by negation if necessary, we can assume the former. Let ψ_i be the conjunction, over all $j \neq i$, of the formulae ϕ_{ij}. Then ψ_i lies in T_i but not in T_j for any $j \neq i$. It follows that, for any $\theta \in T_i$, the sentence

$$(\forall \bar{x})(\psi_i(\bar{x}) \rightarrow \theta(\bar{x}))$$

belongs to the theory Σ. Also, the sentence $(\exists \bar{x})(\psi_i(\bar{x})$ belongs to Σ, and so holds in every model of Σ; but a tuple satisfying ψ_i satisfies every formula in T_i. Thus (i) holds.

Let $\bar{c}$ be an n-tuple of new constant symbols. Then $T \cup \{\theta(\bar{c}) : \theta \in T_i\}$ is a complete consistent theory, which also satisfies the hypotheses of (2.17). Thus every 1-type over this theory is realised in any one of its models. It follows that

($\Diamond$) *If M, N are models of Σ, $\bar{a}$ and $\bar{b}$ are tuples in M and N realising the same type, and x a point of M, then there is a point y of N such that $(\bar{a}, x)$ and $(\bar{b}, y)$ realise the same type.*

It was explained in §2.5 that conclusions (ii) and (iii) of (2.17) follow from ($\Diamond$) by the back-and-forth argument.

The proof of (2.18), like that of Gödel's completeness theorem, involves constructing the required models of Σ. I will not give it here.

Now I turn to the proof of Fraïssé's theorem, which depends on the following preliminary result:

(2.19) *the countable structure M is homogeneous if and only if the following holds:*

($\spadesuit$) *For every A, $B \in$ Age(M) with $A \subseteq B$ and $|B| = |A| + 1$, every embedding of A into M can be extended to an embedding of B into M.*

Proof. Suppose first that M is homogeneous; let A and B be as in the statement of ($\spadesuit$), and $f : A \rightarrow M$ an embedding. There is an embedding $g : B \rightarrow M$, since $B \in$ Age(M). Now $g^{-1}f$ is an isomorphism from Ag to Af; by homogeneity, it extends to an automorphism h of M. Then $gh : B \rightarrow M$ is an embedding extending f.

Conversely, suppose that ($\spadesuit$) holds in the two countable structures M and N with the same age. Then we have:

($\heartsuit$) *Let $\bar{a}$, $\bar{b}$ be n-tuples of distinct elements of M, N respectively, carrying isomorphic substructures, and x a point of M. Then there is a point y of N such that $(\bar{a}, x)$ and $(\bar{b}, y)$ carry isomorphic substructures.*

Once again, this is just what is required to apply back-and-forth. Using this, we conclude that M and N are isomorphic, and that M is homogeneous. In particular, the uniqueness part (2.13) of Fraïssé's theorem is proved.

Now we can prove (2.12). First, the necessity of the conditions. Clearly, the age of any countable structure is closed under isomorphism and under taking substructures, and contains only countably many non-isomorphic members. We must show that the age of a homogeneous structure M has the amalgamation property. Let A, B_1, B_2, f_1, f_2 be as in the statement of the AP. We may assume that A, B_1, $B_2 \subseteq M$, and (by homogeneity) that $A \subseteq B_1$ with f_1 the identity map on A. Now the isomorphism f_2 between finite substructures of M extends to an automorphism h of M. Take $C = B_1 h \cup B_2$, with g_1 the restriction of h to B_1, and g_2 the identity on B_2.

Suppose conversely that C satisfies the four hypotheses. We build a countable structure M satisfying ($\spadesuit$) whose age is C in stages, as follows.

First we list all pairs (A, B) of structures in C for which $A \subseteq B$ and $|B| = |A| + 1$ (up to isomorphism): (A_0, B_0), (A_1, B_1), ...

We begin the inductive construction with $M_0 = \emptyset$.

Suppose that after the nth stage we have built a finite structure $M_n \in C$, with $M_{n-1} \subseteq M_n$. We aim, in the next stage, to construct $M_{n+1} \in C$ such that every embedding $A_i \to M_n$ ($i \le n$) can be lifted to an embedding $B_i \to M_{n+1}$. This is done by listing the finitely many embeddings $A_i \to M_n$, and applying the amalgamation property to each in turn.

Now the union of the structures M_n is a structure M which satisfies ($\spadesuit$), and so is homogeneous. Moreover, its age is C. For every non-empty member of C occurs as B_n for some n, and so is a substructure of M; conversely, every finite substructure of M is contained in M_n for some n, and so is in C (since $M_n \in C$ and C is closed under taking substructures).

Remark. By the Ryll-Nardzewski theorem, a homogeneous structure M which has only finitely many n-element substructures for each n is $\aleph_0$-categorical. But it is easy now to see this directly. A system of axioms is as follows:

(H1) For each structure in Age(M), the (existential) sentence asserting that some n points carry a structure isomorphic to it. (This is not strictly necessary, since these axioms are implied by those in (H3) below.)

(H2) For each n, the (universal) sentence saying that any n points carry a structure isomorphic to one of the (finitely many) n-element structures in Age(M).

(H3) For each pair (A, B) of structures in Age(M) satisfying the hypotheses of (♠), the (universal-existential) sentence asserting that the instance of (♠) is valid.

2.9. APPENDIX: QUANTIFIER ELIMINATION AND MODEL COMPLETENESS

The proofs of the theorems of Ryll-Nardzewski and Fraïssé show marked similarities. In fact, the connection goes deeper, as we shall see in this section. The presentation here is based on Angus Macintyre's talks to the Oxford-QMC "Model theory and permutation groups" seminar.

Let M be a structure over a language L. How can M fail to be homogeneous? If $f : X \to Y$ is an isomorphism between finite substructures of M which fails to extend to an automorphism, then X and Y must be somehow distinguished by outside elements. For example, suppose that M is a graph. Then X and Y might have different numbers of common neighbours, or the distances in M between non-adjacent pairs of vertices in X and Y may differ, and so on. These are first-order properties. This suggests that we will get a more detailed hold on the structure by considering, not the isomorphism types, but the first-order types of finite tuples.

Let M and N be two L-structures, and f a map from M to N. We say that f is *elementary* if, for any tuple $\bar{a}$ in M and formula $\phi(\bar{x})$ of L, if $\phi(\bar{a})$ holds in M, then $\phi(\bar{a}f)$ holds in N. Note:

(i) We don't need to say "if and only if" here, since the negation of f is also a formula.

(ii) Since L implicitly includes equality, an elementary map cannot map unequal elements of M to the same element of N; that is, it is one-to-one.

(iii) If ϕ is a sentence (a formula without free variables) true in M, then ϕ is true in N. So Th(M) = Th(N).

If M is a substructure of N and the identity map on M is elementary, we call M an *elementary substructure* of N, and N an *elementary extension* of M.

If M is homogeneous, then all (first-order) properties of a tuple of elements of M are determined by the isomorphism type of the tuple, that is, by properties given by quantifier-free formulae. Accordingly, we say that the complete first-order theory Σ *has quantifier-elimination* if, given any formula $\phi(\bar{x})$, there is a quantifier-free formula $\psi(\bar{x})$ such that

$$(\forall \bar{x})(\phi(\bar{x}) \leftrightarrow \psi(\bar{x}))$$

is provable from Σ. (By the Completeness Theorem (1.4), this is the same as saying that the displayed sentence holds in every model of Σ.) Moreover, M *has quantifier-elimination* if $\mathrm{Th}(M)$ does. Now we have the following result:

(2.20) *A first-order theory Σ has quantifier-elimination if and only if every model of Σ has an elementary extension which is a homogeneous model of Σ.*

Sketch proof. First of all, any first-order structure M has an elementary extension N with the property that, if $\bar{a}$ and $\bar{b}$ are tuples in M with $\mathrm{tp}_M(\bar{a}) = \mathrm{tp}_M(\bar{b})$, then some automorphism of N carries $\bar{a}$ to $\bar{b}$. (This is proved by arguments like those in §2.8; it is not meant to be obvious.) If the theory has quantifier-elimination, then tuples have the same type precisely when they are isomorphic. This shows the necessity of the condition. Conversely, suppose that quantifier elimination fails, so that there are isomorphic tuples in some model M whose types differ. Since types are preserved by elementary extensions, no such extension of M can be homogeneous.

A slight improvement of this result involves the notion of an existentially closed substructure. Recall that an existential formula is one of the form $(\exists \bar{y})\psi(\bar{x}, \bar{y})$, where ψ is quantifier-free. The substructure M of N is *existentially closed* in N if, for every tuple $\bar{a}$ of M and existential formula $\phi(\bar{x})$, $\phi(\bar{a})$ holds in M if and only if it holds in N. (In other words, the identity map from M to N preserves existential formulae. Clearly this is weaker than saying that N is an elementary extension of M.)

(2.21) *A first-order theory Σ has quantifier-elimination if and only if every model of Σ is existentially closed in a homogeneous model of Σ.*

Proof. The necessity follows from (2.20). For the converse, note first that the hypothesis implies a weak form of quantifier elimination: every existential formula is equivalent to a quantifier-free formula. Taking negations, every universal formula is equivalent to a quantifier-free formula. Now the general result follows by induction on the number of quantifiers in the formula.

(2.22) *Let M be countable and $\aleph_0$-categorical. Then M has quantifier elimination if and only if it is homogeneous.*

This is an immediate corollary of (2.20) and the fact that, if M is $\aleph_0$-categorical, then every countable elementary extension of M is isomorphic to M.

A closely related notion is that of model-completeness. A theory Σ is *model-complete* if every embedding of one model of Σ into another is an elementary map. According to a theorem of Abraham Robinson, it suffices that every embedding preserves existential formulae. (The argument is like the deduction of (2.21) from (2.20) above.) Trivially, a theory with quantifier-elimination is model-complete. The converse is false: but Σ has quantifier-elimination if and only if it is model-complete and the class of all substructures of all models of Σ has the amalgamation property.

Now we have the following result:

(2.23) *Let M be countable and $\aleph_0$-categorical. Then $\mathrm{Th}(M)$ is model-complete if and only if it is an $(\forall\exists)$-theory, that is, equivalent to a class of $(\forall\exists)$-sentences.*

We saw at the end of the last section that, if M is countable, homogeneous and $\aleph_0$-categorical, then $\mathrm{Th}(M)$ is an $(\forall\exists)$-theory. So finally:

(2.24) *A countable homogeneous $\aleph_0$-categorical structure is model-complete.*

Example. The theory of $\mathbf{Q} \cap [0, 1]$ (as totally ordered set) is not model-complete; the embedding of this model in $\mathbf{Q} \cap [-1, 1]$ is not elementary. It can be made model-complete (and homogeneous) by adding to the language two constant symbols to stand for the least and greatest elements.

Exercise

(i) For finite structures, any elementary map is an isomorphism.

(ii) Show that a finite structure has quantifier-elimination if and only if it is homogeneous.

(iii) Can you prove this without using the general theorems of this section?

2.10. APPENDIX: THE RANDOM GRAPH

In this section, R denotes the unique countable homogeneous universal graph (the structure we called R_1 in §2.6). R is called the *random graph*, for reasons that will become clearer in Chapter 4.

(2.25) *R is the unique countable graph with the property:*

(♣) *For any finite set A of vertices, and any subset B of A, there is a vertex z whose neighbours in A are precisely the vertices of B.*

Proof. Condition (♣) is just a restatement of (♠) of §2.8 for the case of a graph whose age consists of all finite graphs. (Check that a graph satisfying (♣) really does contain all finite graphs.)

The graph R has many interesting properties. I will prove just one of them. Several more are given in the Exercises below. Others are deferred until Chapter 4.

(2.26) *The group* $\mathrm{Aut}^*(R)$ *of automorphisms and anti-automorphisms of R is 2-transitive and contains* $\mathrm{Aut}(R)$ *as a subgroup of index 2.*

[An *anti-automorphism* of a graph is a permutation of the vertex set which maps edges to non-edges and non-edges to edges, i.e. interchanges the graph with its complement.]

First note that R is isomorphic to its complement $\bar{R}$ — for condition (♣) is self-complementary, so $\bar{R}$ is also a countable graph satisfying (♣). Now an isomorphism from R to $\bar{R}$ is an anti-automorphism of R. There is a homomorphism from $\mathrm{Aut}^*(R)$ to $\mathbf{Z}/(2)$ mapping automorphisms to $\bar{0}$ and anti-automorphisms to $\bar{1}$; our remarks imply that it is onto, so that $\mathrm{Aut}(R)$ has index 2 in $\mathrm{Aut}^*(R)$. Finally, let α and β be adjacent vertices of R. If γ and δ are adjacent, then some automorphism maps (α, β) to (γ, δ), by homogeneity. Suppose that γ and δ are non-adjacent. Then an arbitrary anti-automorphism maps (α, β) to some non-adjacent pair (ϵ, ζ). and then there is an automorphism carrying (ϵ, ζ) to (γ, δ). So $\mathrm{Aut}^*(R)$ is 2-transitive, as claimed.

Note. $\mathrm{Aut}^*(R)$ is not 3-transitive; triples containing 0 or 3 edges of R are not equivalent to triples containing 1 or 2 edges.

Exercises

1. Show that each of the following graphs is isomorphic to R.

(a) The vertex set is $\mathbf{N}$. For $x < y$, x and y are adjacent if and only if 2^x occurs in the unique expression for y as a sum of distinct powers of 2 (i.e. in the base 2 expansion of y.)

(*Subsidiary exercise.* Write $x \in y$ if $x < y$ and the above relation between x and y holds. Show that we obtain a model for the Zermelo-Fraenkel axioms for set theory, with the exception of the axiom of infinity.)

(b) The vertices are the primes congruent to 1 modulo 4. Vertices p and q are adjacent if and only if p is a quadratic residue mod q. (The Law of Quadratic Reciprocity shows that the graph is undirected. To prove that it is isomorphic to R, you will need the Chinese Remainder Theorem and Dirichlet's Theorem on primes in arithmetic progression.)

2. Show that, in either of the presentations of R in Exercise 1, the group of all recursive functions which are automorphisms of R acts homogeneously on R. Is the same true with "primitive recursive" in place of "recursive"?

Research problem. What is the structure of these (countable) groups? How, if at all, does the choice of recursive presentation of R affect the groups? Generalise to other $\aleph_0$-categorical or homogeneous structures.

3. A graph Γ_1 is a *spanning subgraph* of Γ_2 if they have the same set of vertices, and the edge set of Γ_1 is a subset of that of Γ_2. Show that a graph Γ with countably many vertices is isomorphic to a spanning subgraph of R if and only if it satisfies the following condition: *for any finite set C of vertices of Γ, there is a vertex z adjacent to no vertex in C.* (This is condition (♣) under the restriction that $B = \emptyset$.)

Deduce that R has one- and two-way infinite spanning paths.

Remark. Spanning subgraphs are in a sense dual to induced subgraphs. By Exercise 4 of §2.6, every finite or countable graph is isomorphic to an induced subgraph of R.

4. Prove that R contains a maximal complete subgraph. (Of course, use of the Axiom of Choice is forbidden!)

5. Consider the topology on the vertex set of R in which a basis for the open sets consists of the sets

$$N(U) = \{z : z \sim u \text{ for all } u \in U\}$$

for all finite sets U of vertices. Show that this topology is homeomorphic to $\mathbf{Q}$ (with the usual topology). (You will probably need the characterisation of $\mathbf{Q}$ as a countable, totally disconnected, T_1-space without isolated points by Sierpiński (1920) — see also Neumann (1985a).)

Deduce that Aut(R) is a subgroup of the group Homeo($\mathbf{Q}$) of homeomorphisms of $\mathbf{Q}$.

(We saw that Homeo($\mathbf{Q}$) is highly transitive in Exercise 3 of §2.4.)

(Harder.) Is the group Aut$^*(R)$ embeddable in Homeo($\mathbf{Q}$)?

Remark. For more on embeddability of permutation groups in Homeo($\mathbf{Q}$), see Mekler (1986) and Truss (1986).

6. A permutation g of the vertex set of R is a *switching automorphism* if Rg can be obtained from R by the following operation (called *switching*):

Choose a set Y of vertices. Delete all edges between Y and its complement, and add in edges joining points of Y to points of its complement which were previously not joined; leave all other edges and non-edges unaltered.

Show that the switching automorphisms of R form a group G, which is 2-transitive but not 3-transitive, and the stabiliser of a point (in G) is isomorphic (as permutation group) to Aut(R). (Of course, G also contains a transitive subgroup isomorphic to Aut(R), since every automorphism is a switching automorphism.)

Show also that the group generated by the switching automorphisms and the anti-automorphisms is 3-transitive but not 4-transitive.

Remarks. (1) See Seidel (1976) for more on switching of finite graphs, a concept which has connections with topological coverings, Euclidean line systems, and finite simple groups.

(2) A recent result of Dugald Macpherson and Simon Thomas (to appear) asserts that the only closed proper supergroups of Aut(R) are Aut$^*(R)$, the two groups of this Exercise, and the symmetric group.

7. A permutation of the vertex set of R is *almost an automorphism* if it maps only finitely many edges onto non-edges and *vice versa*. Show that the almost-automorphisms of R form a highly transitive permutation group.

(See Truss (1989) for more about this group.)

Problem. Which other graphs have interesting groups of almost-automorphisms?

8. Let A be a finite set of vertices of R, and $B \subseteq A$. Show that
$$\{v : \textit{the set of neighbours of } v \textit{ in } A \textit{ is precisely } B\}$$
carries a subgraph isomorphic to R.

Deduce that, if the vertex set of R is partitioned into finitely many parts, then one of these parts contains an induced copy of R.

Problem. For what other "nice" structures does this "pigeonhole principle" hold?

3

Examples and growth rates

3. Examples and growth rates

3.1. MONOTONICITY

In this chapter, I consider the problem of determining which sequences of natural numbers can occur as (f_n) or (F_n) for some oligomorphic permutation group. The problem in this generality is quite out of reach. All I can do is to illustrate by examples some of the possibilities that can occur, and to describe some of the restrictions on realisable sequences which have been established.

We saw in the last chapter that any permutation group is a dense subgroup of the automorphism group of a homogeneous relational structure, so it suffices to consider these groups. Moreover, since the group is oligomorphic, the structure is $\aleph_0$-categorical. The results of §2.5 imply that the sequences (f_n) and (F_n) realised by oligomorphic groups are precisely those which enumerate unlabelled and labelled structures respectively in a class of finite structures satisfying Fraïssé's hypotheses (notably, the amalgamation property). From this point of view, two subcases commend themselves to us:
  (a) classes having the strong amalgamation property;
  (b) ages of homogeneous structures over finite relational languages.
Other subcases are obtained by imposing model-theoretic conditions in the neighbourhood of stability on the homogeneous structure.

The most important basic restriction on realisable sequences is that they must be monotonic. For (F_n), this is a triviality:

(3.1) $F_{n+1} \geq F_n$, with equality if and only if $F_{n+1} = 1$.

For there is a surjection from $(n+1)$-tuples to n-tuples (of distinct elements) defined by rubbing out the last entry; it is a morphism of G-spaces (i.e. it satisfies the commutative diagram of §1.2), and so it induces a map on orbits, which is also a surjection. Hence the inequality. If equality holds, then this map is a bijection, and so each orbit on n-tuples extends uniquely to an orbit on $(n+1)$-tuples. This implies that the stabiliser of any n-tuple acts transitively on the remaining points. It is straightforward to show that this implies $(n+1)$-transitivity. The converse is clear.

From now on, I shall consider only the sequence (f_n). The analogue of (3.1) is much less trivial, and a characterisation of equality is lacking.

(3.2) $f_{n+1} \geq f_n$.

I will give (in outline) two proofs of this result, using quite different ideas. Both of these have been useful in other situations.

First proof (linear algebra). Let V_n denote the vector space of functions from the set of n-subsets of Ω to $\mathbf{Q}$ (any field of characteristic 0 would do). With pointwise addition and scalar multiplication, V_n is a vector space over $\mathbf{Q}$. Now define a map $\theta : V_n \to V_{n+1}$ by the rule

$$(f\theta)(A) = \sum_{a \in A} f(A \setminus \{a\}),$$

where $f \in V_n$, $|A| = n+1$.

This map θ is obviously linear. The technical part of the proof consists in showing that it is a monomorphism. Accordingly, suppose that $f \in \ker(\theta)$ and $f(B) \neq 0$ for some n-set B. Let Γ be any (possibly finite) set containing B. Let $V_n[\Gamma]$ be the vector space of rational-valued functions on the set of n-subsets of Γ, $f[\Gamma]$ the restriction of f to the set of n-subsets of Γ (an element of $V_n[\Gamma]$), and $\theta[\Gamma]$ the map from $V_n[\Gamma]$ to $V_{n+1}[\Gamma]$ defined in the same way as θ. Then $f[\Gamma]$ belongs to the kernel of $\theta[\Gamma]$, and is non-zero (it doesn't vanish on B). So, to prove the assumption, it suffices to demonstrate it in the case when Ω is finite and sufficiently large.

The problem is now one of finite combinatorics. It turns out that "sufficiently large" means $|\Omega| \geq 2n+1$. (This is best possible, since if $|\Omega| = m \leq 2n$, then

$$\dim(V_{n+1}) = \binom{m}{n+1} < \binom{m}{n} = \dim(V_n),$$

and no monomorphism from V_n to V_{n+1} can exist.)

So assume that $|\Omega| = 2n+1$. There is a bijection between $(n+1)$-subsets of Ω and the complementary n-sets. Using it, we can rewrite θ as a map from V_n to V_n as follows:

$$(f\theta)(A) = \sum \{f(B) : |B| = n, B \cap A = \emptyset\},$$

where A is an n-set. Now suppose that $f\theta = 0$. Let J be a j-set, where $j < n$. Throughout the argument below, A and B denote n-sets. Let $x_J = \sum\{f(B) : B \cap J = \emptyset\}$. Then

$$0 = \sum\{\sum\{f(B) : B \cap A = \emptyset\} : A \supseteq J\}$$
$$= \sum\{\sum\{f(B) : A \supseteq J, A \cap B = \emptyset\} : B \cap J = \emptyset\}$$
$$= (n+1-j)x_J,$$

so $x_J = 0$.

Now let $|I| = i$, $|J| = j$ with $I \cap J = \emptyset$, $i + j \le n$. Set

$$y_{I,J} = \sum\{f(B) : B \supseteq I, B \cap J = \emptyset\}.$$

Then $y_{\emptyset,J} = x_J = 0$; also, for $i \in I$,

$$y_{I,J} = y_{I\setminus\{i\},J} - y_{I\setminus\{i\},J\cup\{i\}}$$

(by considering the sets B contributing to the sums); so

$$y_{I,J} = \emptyset$$

(by induction on i). Hence

$$f(A) = y_{A,\emptyset} = 0$$

for any n-set A.

Now we return to the proof of (3.2). Let G be a permutation group on Ω. Then G acts as a linear group on V_n and V_{n+1}, and these actions are intertwined by θ. So, if V_n^G denotes the set of G-invariant functions in V_n, then θ maps V_n^G to V_{n+1}^G (and is still a monomorphism). Assuming that f_n is finite, V_n^G is spanned by the characteristic functions of the G-orbits, and so its dimension is f_n. The result follows.

Second proof (Ramsey's theorem). We use the strengthening of Ramsey's theorem given as (1.10).

Let G act on Ω with $r = f_n$ orbits on n-subsets. We define a colouring of the n-sets with r colours by associating the ith colour with the sets in the ith orbit. Assume the colours have been ordered as described in the conclusion of (1.10), and let $X_1, \ldots, X_r$ be the infinite sets guaranteed by that result. For each i, let B_i be an $(n+1)$-set contained in X_i and containing an n-set with the ith colour. Then $B_1, \ldots, B_r$ are coloured with distinct sets of colours. Since G preserves the colouring, $B_1, \ldots, B_r$ lie in distinct orbits; so $f_{n+1} \ge r = f_n$.

The argument proves a little more:

(3.3) *The number of orbits of G on countable subsets of Ω with countable complements is at least f_n.*

(For the sets $X_1, \ldots, X_r$ of the proof can be assumed to have these properties and lie in distinct G-orbits.)

The two obvious questions raised by (3.2) are:
(i) What happens if $f_n = f_{n+1}$?
(ii) How fast does the sequence (f_n) grow?

We return to these questions in §§3.4–3.6, after describing (in the next two sections) some examples.

3.2. DIRECT AND WREATH PRODUCTS

We begin with a relation between the numbers of orbits of a group and a subgroup.

(3.4) *If $G \leq H$, then*
(i) $f_n(G) \geq f_n(H)$ *and* $F_n(G) \geq F_n(H)$;
(ii) $f_{n+1}(G) - f_n(G) \geq f_{n+1}(H) - f_n(H)$.

The first part requires no comment. For the second, use the terminology of the linear algebra proof of (3.2). We have

$$V_n^G \theta \cap V_{n+1}^H \subseteq V_n^H \theta,$$

for an element of the left-hand side is the image under θ of a unique element of V_n, which is necessarily fixed by H. So

$$
\begin{aligned}
f_{n+1}(H) - f_n(H) &= \dim(V_{n+1}^H / V_n^H \theta) \\
&\leq \dim(V_{n+1}^H / (V_n^G \theta \cap V_{n+1}^H)) \\
&= \dim((V_{n+1}^H + V_n^G \theta) / V_n^G \theta) \\
&\leq \dim(V_{n+1}^G / V_n^G \theta) \\
&= f_{n+1}(G) - f_n(G).
\end{aligned}
$$

This result is relevant to both the questions posed in the last section. For the growth rate of the sequence $(f_n(G))$ is at least as great as that for any supergroup of G; and, if $f_n = f_{n+1}$ holds for G, then it holds for any supergroup of G. So little is lost by passing to supergroups.

This is useful in considering intransitive or imprimitive groups. For, as we saw in §1.2, these groups are contained in maximal intransitive or imprimitive groups which are cartesian or wreath products; and these turn out to be easier to analyse in terms of their "building blocks".

Certain generating functions are useful for expressing these results. We have already met the exponential generating function

$$F_G(t) = \sum_{n \geq 0} F_n(G)t^n/n!$$

for the sequence (F_n). We use the *ordinary generating function* (o.g.f.)

$$f_G(t) = \sum_{n \geq 0} f_n t^n$$

for (f_n).

(3.5) *(i)* $f_{G \times H}(t) = f_G(t) \cdot f_H(t)$;
(ii) $F_{G \times H}(t) = F_G(t) \cdot F_H(t)$.

(The intransitive action of the direct product is used here.)

Thus, for example, if H is highly homogeneous, then $(f_n(G \times H))$ is the sequence of partial sums of $(f_n(G))$. In particular, if S is the symmetric group, then

$$f_n(S^r) = \binom{n + r - 1}{r - 1},$$

which is a polynomial of degree $r - 1$ in n with leading coefficient $1/(r-1)!$.

(3.6) $F_{G \,\mathrm{Wr}\, H}(t) = F_H(F_G(t) - 1)$.

In contrast to (3.5), it is not possible to calculate the sequence $(f_n(G \,\mathrm{Wr}\, H))$ from $(f_n(G))$ and $(f_n(H))$ alone. The extra information necessary is the so-called "modified cycle index" of H, and is discussed in §3.7. I will anticipate this by mentioning two special cases. Let S denote the symmetric group, and A the highly homogeneous group $\mathrm{Aut}(\mathbf{Q}, <)$.

(3.7) *(i)* $f_{G \,\mathrm{Wr}\, S}(t) = \prod_{n \geq 1}(1 - t^n)^{-f_n(G)}$;
(ii) $f_{G \,\mathrm{Wr}\, A}(t) = (2 - f_G(t))^{-1}$.

Examples. Let C_2 be the cyclic group of order 2, acting regularly. Then $f_{C_2}(t) = 1 + t + t^2$.

(i) $f_n(C_2 \operatorname{Wr} S) = 1 + \lfloor \frac{1}{2}n \rfloor$. (The generating function is $(1-t)^{-1}(1-t^2)^{-1}$.) Note that $f_{2n} = f_{2n+1}$ for all n.

(ii) $f_n(C_2 \operatorname{Wr} A)$ is the nth Fibonacci number. (The generating function here is $(1-t-t^2)^{-1}$.) This is roughly τ^n, where $\tau = \frac{1}{2}(1+\sqrt{5})$ is the golden ratio.

In the same vein, $f_n(S \operatorname{Wr} S) = p(n)$, the number of partitions of the integer n (with generating function $\prod_{n \geq 1}(1-t^n)^{-1}$). Very good asymptotic estimates for $p(n)$ are known; but I'll just observe here that $p(n)$ is roughly $\exp(n^{\frac{1}{2}})$. (This is not a valid asymptotic expansion; but $\log p(n) \sim n^{\frac{1}{2}}$.)

For the iterated wreath product of r copies of A, we have $f_n = r^{n-1}$.

Many more interesting examples in the same vein occur in Cameron (1987b).

Result (3.7)(i) can be used to give lower bounds for the growth rates for imprimitive groups G having a congruence with infinitely many classes. For, if G is such a group, and H the group induced on a congruence class by its setwise stabiliser, then $G \leq H \operatorname{Wr} S$, and (3.5)(i) applies. Some fairly precise results are known about this situation. I'll state them in an impressionistic way.

(3.8) *Let $G = H \operatorname{Wr} S$, where S is the symmetric group.*
(i) If $(f_n(H))$ grows like a polynomial of degree r in n, then $f_n(G)$ is roughly $\exp(n^{(r+1)/(r+2)})$.
(ii) If $(f_n(H))$ grows faster than any polynomial, then $(f_n(G))$ grows faster than $\exp(n^c)$ *for any $c < 1$.*
(iii) The radii of convergence of $f_H(t)$ and $f_G(t)$ are equal.

Part (i) gives a variety of possible growth rates faster than any polynomial but slower than exponential. Part (ii) shows, for example, that for the iterated wreath product of any number (greater than two) of symmetric groups, the growth rate is faster than fractional exponential, while part (iii) shows that it is slower than any exponential c^n ($c > 1$).

Exercises

1. Prove some of the assertions in this section.

2. The fact that $f_n(S \operatorname{Wr} S) = p(n)$ is easily seen directly. The set Ω is the union of infinitely many infinite congruence classes; so, with every n-set is associated a partition of n given by the cardinalities of its intersections with the classes. Two n-sets lie in the same orbit if and only if the same partition is associated with each.

Find similar direct proofs of the identifications given above for each of the groups $f_n(C_2 \operatorname{Wr} S)$, $f_n(C_2 \operatorname{Wr} A)$, and $f_n(A \operatorname{Wr} A)$.

3. Let S_r be the symmetric group of finite degree r. Show that
$$f_n(S_r \operatorname{Wr} S) = f_n(S \operatorname{Wr} S_r)$$
for all n. Show that this quantity is roughly a polynomial of degree $r - 1$ in n, and find its leading coefficient.

3.3. SOME PRIMITIVE GROUPS

In this section, I describe (mostly without detailed proof) a few examples of growth rates realised by primitive groups.

Primitive groups cannot be built up from simpler ingredients. We must either describe the group (or a structure whose automorphism group it is) directly, or give a class of finite structures satisfying Fraïssé's conditions. The latter approach usually makes calculation of the sequence (f_n) easier, but it leaves the group somewhat mysterious. More of that later ...

Example 1: Local orders. A *tournament* is a binary relation R such that exactly one of $R(x, y)$, $x = y$, $R(y, x)$ holds for each pair x, y of points (the *trichotomy*). A linear order can be described as a tournament containing no 3-cycles. A *local order* is a tournament with the property that no 4-point substructure consists of a 3-cycle dominating or dominated by a point. In other words, for any x, the sets $\{y : R(x, y)\}$ and $\{y : R(y, x)\}$ are linearly ordered. Finite local orders satisfy Fraïssé's hypotheses, and so there is a unique countable homogeneous local order L containing all the finite ones.

A remarkable theorem of Lachlan (1984) asserts that there are just three countable homogeneous tournaments: as well as L, there is the homogeneous linear order $\mathbf{Q}$, and the "universal" tournament (whose age consists of all finite tournaments).

There is a simple, more-or-less explicit, description of L. From the set of complex roots of unity, select one of each pair $\{\omega, -\omega\}$ in such a way that the points selected are dense on the unit circle. (As a trailer for what comes later, I mention that if the choices are made at random, independently with probability $\frac{1}{2}$ for each outcome, then the resulting set is almost surely dense.) Now take the selected points as the points of the tournament; for $x \neq y$, let $R(x, y)$ hold if and only if the shortest path on the unit circle from x to y is anticlockwise. (Imagine driving round a roundabout, but not in the United Kingdom!) To check the correctness of this construction, it suffices to show that the age of this tournament consists of all finite local orders and that condition ($\spadesuit$) of §2.6 holds.

Now $f_n(\mathrm{Aut}(L))$, which is the number of n-point local orders (up to isomorphism), is asymptotic to $2^{n-1}/n$. (The exact formula is

$$f_n(\mathrm{Aut}(L)) = \frac{1}{2n} \sum_{\substack{d|n \\ d \text{ odd}}} \phi(d) 2^{n/d}$$

where ϕ is Euler's totient function. This sequence also enumerates the output sequences of a shift register of length n, where the bit shifted in is the complement of the bit shifted out.) This number is roughly halved if we use the larger group of automorphisms and anti-automorphisms (orientation reversing permutations) of L. The latter group has the slowest known growth rate of a primitive but not highly transitive group.

A modification gives primitive groups with growth rate roughly r^n for any natural number r. For this, we follow the roundabout construction, but choosing a dense set containing one element from each set of r equally spaced roots of unity; we partition the ordered pairs of distinct elements into r relations $R_0, \ldots, R_{r-1}$, where (x, y) satisfies R_j if and only if the angle (in the positive sense) between x and y lies between $2\pi j/r$ and $2\pi(j + 1)/r$.

Example 2. For the boron trees of §2.6, f_n is roughly c^n, where $c = 2.483\ldots$. (Many more decimal places of c, and a much more precise asymptotic formula for f_n, are known. Comtet (1974) gives a detailed account of this; a sketch of the calculation of c is given in §3.7.)

Replacing the group in question by the stabiliser of a point doesn't change significantly the approximation to f_n, and hence the value of c (see Exercise 2). For the latter group, the numbers f_n are the *Wedderburn-Etherington numbers*, which count rooted binary trees where no distinction is made between the left-hand and right-hand branches at any node. They satisfy the recurrence relation

$$f_n = f_1 f_{n-1} + f_2 f_{n-2} + \ldots,$$

where the last term in the sum is given by

$$
\begin{cases}
f_{(n-1)/2} f_{(n+1)/2}, & \text{if } n \text{ is odd;} \\
\binom{f_{n/2}+1}{2}, & \text{if } n \text{ is even.}
\end{cases}
$$

From this, it follows that the o.g.f. is $f(t) = 1 + x(t)$, where

$$
x(t) = t + \frac{1}{2}(x(t)^2 + x(t^2));
$$

c is the reciprocal of its radius of convergence.

If the left-right distinction is made, as usual in computer science and elsewhere, the resulting numbers are called the *Catalan numbers*; they satisfy the recurrence

$$
f_n = f_1 f_{n-1} + f_2 f_{n-2} + \ldots + f_{n-1} f_1;
$$

explicitly,

$$
f_n = \binom{2n-2}{n-1} / n,
$$

which is "roughly" 4^n. This sequence is also realised by a group. There are a large number of variations on these constructions, which give groups for which the growth rate is roughly exponential. See Cameron (1987b).

The third example depends on a very simple but powerful construction technique.

(3.9) *Let C and D be classes of finite structures satisfying Fraïssé's hypotheses and having the strong amalgamation property. Let $C \wedge D$ denote the class of structures consisting of a finite set of points carrying both a C-structure and a D-structure (with no relation between them). Then $C \wedge D$ also satisfies Fraïssé's conditions and has the strong amalgamation property.*

Example 3. Consider $C \wedge C$, where C is the class of linear orders. There are $n!$ n-element structures in this class, since the second ordering is an arbitrary permutation of the first. Thus $f_n = n!$, growing a little faster than exponentially (but slower than $\exp(n^c)$ for any $c > 1$). A slight reduction (roughly by a factor of 8) is obtained by permitting either order to be reversed or the two orders to be interchanged.

Example 4. Consider $G = \mathrm{Sym}(\mathbf{N})$, acting on the set Ω of 2-element subsets of $\mathbf{N}$. An n-element subset of Ω is the edge set of a graph with vertex set $\mathbf{N}$ having n edges. Removing vertices lying on no edge, we see that $f_n(G)$ is equal to the number of graphs with n edges and no isolated vertices, up to isomorphism. This number grows faster than exponentially. (If m is close to $n/\log n$, then most graphs with m vertices and n edges have no isolated vertices, and there are at least $\binom{\frac{1}{2}m(m-1)}{n}/m!$ such graphs up to isomorphism.) A good asymptotic estimate seems not to be known, despite detailed studies of related problems by Wright (1980).

A very closely related example is the group $G = \mathrm{Sym}(\mathbf{N})\,\mathrm{Wr}\,\mathrm{Sym}(2)$ in its product action. For this group, $f_n(G)$ is the number of bipartite graphs with n edges and no isolated vertices, with a prescribed bipartition. (A connected bipartite graph has a unique bipartition; a bipartite graph with d connected components has 2^{d-1} bipartitions.) Again, detailed asymptotics are unknown, but the growth appears to be a little faster than exponential.

Example 5. For the random graph (the universal homogeneous graph R of §§2.6, 2.10), f_n is the number of n-vertex graphs, which is asymptotically $2^{\frac{1}{2}n(n-1)}/n!$. This is roughly $2^{P(n)}$, where P is a polynomial of degree 2 with leading coefficient $\frac{1}{2}$. The same asymptotic estimate holds for the universal homogeneous tournament (see Example 1), and similar estimates (with P replaced by a polynomial of degree $k+1$) for the universal hypergraph R_k.

Example 6. For the collineation group of a projective or affine space over $\mathrm{GF}(q)$, f_n is roughly $q^{\frac{1}{4}n^2}$. (Cameron and Taylor (1985) give the formula

$$F_n^*(G) = \sum_{k=0}^{n} \begin{bmatrix} n \\ k \end{bmatrix}_q \phi_k(G)$$

for an infinite-dimensional linear group G over $\mathrm{GF}(q)$, acting on the set of vectors, where $\phi_k(G)$ is the number of orbits of G on linearly independent k-tuples of vectors, and $\begin{bmatrix} n \\ k \end{bmatrix}_q$ is the Gaussian coefficient, the number of k-dimensional subspaces of an n-dimensional vector space over $\mathrm{GF}(q)$. So, if G is the general linear group, then $\phi_k(G) = 1$ for all k, and so $F_n^*(G)$ is the total number of subspaces of $\mathrm{GF}(q)^n$; this number is about $q^{\frac{1}{4}n^2}$. From this, a formula for $F_n(G)$, and hence upper and lower bounds for $f_n(G)$, are obtained from (2.3) and (2.1). Finally, the value of F_{n+1} for the full affine group is equal to that of F_n for the general linear group acting on the non-zero vectors.)

Exercises

1. Use the formula of Cameron and Taylor to calculate F_n for the symplectic group of countable dimension over $\mathrm{GF}(q)$. (To apply the formula, it is necessary to calculate the number of orbits of this group on linearly independent k-tuples; show that this is equal to the number of skew-symmetric $k \times k$ matrices, using Witt's lemma.)

2. Let H be the stabiliser of a point in G. Show that

$$f_{n+1}(G) \le f_n(H) \le (n+1)f_{n+1}(G)$$

and deduce that the existence and value of $\lim_{n\to\infty} f_n^{1/n}$ are the same in H as in G. In consequence, if G is closed and H has index less than $2^{\aleph_0}$ in G, then

$$\lim_{n\to\infty} f_n(H)^{1/n} = \lim_{n\to\infty} f_n(G)^{1/n}$$

(if these limits exist; limits superior in general), so that the radii of convergence of $f_G(t)$ and $f_H(t)$ are equal.

Research problems. 1. Does $\lim_{n\to\infty} f_n(G)^{1/n}$ always exist?

2. Let C be the set of limits as in Problem 1 (or limits superior, if the answer is negative — that is, reciprocals of radii of convergence of o.g.f.s). Investigate C. For example, is it uncountable? What is its smallest member apart from 1? What is its smallest limit point? What other gaps occur? What happens when we consider only primitive groups?

3. Find good asymptotic estimates for $f_n(G)$, where G is one of the groups of Example 4.

3.4. HOMOGENEITY AND TRANSITIVITY

Highly homogeneous permutation groups obviously hold a special place with respect to both the questions of §3.1, viz. groups with $f_n = f_{n+1}$ (they satisfy this for all n), and rate of growth of (f_n) (no growth at all!) They are special to me as well, as (3.10) below was my introduction to this subject. These permutation groups might be called "monomorphic", if this word did not have an established usage already. (There is a precedent; see Frasnay (1974).)

We have already observed that the group $A = \mathrm{Aut}(\mathbf{Q}, <)$ is highly homogeneous but not highly transitive (not even 2-transitive). It has a 2-transitive supergroup B, the group of permutations of $\mathbf{Q}$ which preserve or reverse the order; however, B is not 3-transitive, since it preserves the ternary "betweenness" relation on $\mathbf{Q}$. Another 2-transitive but not 3-transitive highly homogeneous group is the group C of permutations of the complex roots of unity which preserve the cyclic order; and the group D of permutations which preserve or reverse the cyclic order is 3-transitive but not 4-transitive. Finally, of course, the symmetric group S is highly transitive. This is essentially all:

(3.10) *A highly homogeneous permutation group of countable degree is a dense subgroup of one of A, B, C, D or S.*

In other words, there are exactly five closed highly homogeneous groups of countable degree. Moreover, the technique of §2.1 shows that a highly homogeneous but not highly transitive group of any infinite degree can be no more than 3-transitive, and preserves or reverses a linear or circular order.)

With any highly homogeneous group G is associated a sequence (G_n) of finite permutation groups, where G_n is the group induced on an n-set by its setwise stabiliser. (Clearly the choice of n-set is irrelevant.) My proof (Cameron (1976)) of (3.10) involved careful consideration of these finite groups, with a good deal of case analysis. A completely different, more conceptual proof was given independently by Higman (1977) and Hodges (using results of Pouzet (1976), Frasnay (1974) and Hodges, Lachlan & Shelah (1977)). I give a brief outline.

Suppose that G is highly homogeneous but not n-transitive, and let R be a G-orbit on n-tuples of distinct elements, regarded as an n-ary relation. Then $G \leq \mathrm{Aut}(\Omega, R)$, so the latter group is also highly homogeneous but not n-transitive, and we may assume that equality holds here.

The relation R on Ω is said to be *chainable* if there is a linear order $<$ on Ω such that the unique order-preserving bijection between two finite sets of the same size also preserves R. (For the kind of relation we have here, which holds on an n-tuple only if its arguments are all distinct, there is an equivalent formulation: R can be specified by giving a set S of permutations of $\{0, \ldots, n-1\}$, by the rule that, for any $\alpha_0, \ldots, \alpha_{n-1} \in \Omega$ with $\alpha_0 < \ldots < \alpha_{n-1}$, and any $\pi \in \mathrm{Sym}(n)$,

$$(\alpha_{0\pi}, \ldots, \alpha_{(n-1)\pi}) \in R \quad \Longleftrightarrow \quad \pi \in S.$$

By Ramsey's theorem, R is chainable on an infinite subset of Ω. (For take any linear order $<$ on Ω. Then, given any n-set $A = \{\alpha_0, \ldots, \alpha_{n-1}\}$ with $\alpha_0 < \ldots < \alpha_{n-1}$, there is a subset S of $\mathrm{Sym}(n)$ such that the restriction of R to A is given by the displayed rule in the preceding paragraph. Now there are only finitely many subsets of $\mathrm{Sym}(n)$; so, by Ramsey's theorem, there is an infinite subset Δ of Ω for which all n-sets give rise to the same subset S of $\mathrm{Sym}(n)$.)

Now a compactness argument shows that R is chainable on all of Ω. (Any finite subset of Ω is equivalent, under $\mathrm{Aut}(R)$, to a subset of Δ, and hence the restriction of R to it is chainable. Now consider the language L with two relation symbols $<$ and R and countably many constant symbols (one for each point of Ω), and the theory saying that $<$ is a linear order, giving all instances of R on Ω, and asserting that R is derived from $<$ (by the given rule) on any n-subset of Ω. The compactness theorem (1.4), with our observation about finite subsets of Ω, shows that this set of sentences is satisfiable.)

Now we know that $\text{Aut}(R)$ permutes the set of all linear orders from which R can be defined by the given rule. It is possible to show (see Higman (1977), Hodges-Lachlan-Shelah (1977)) that this set consists of a single order, or an order and its reverse, or all orders obtained from a given one by "sliding" round a circle, or all these and their reverses, or else that every finite set is ordered in all possible ways by these orders. From this the result follows.

Can this result be quantified? It was already shown in Cameron (1976) that, if m is sufficiently large compared to n, then an m-homogeneous but not n-transitive group is contained in a group isomorphic to one of A, B, C or D. (Since all the others are contained in D, the conclusion can be simplified.) The prototype of such results, due to MacDermott (personal communication), is easy:

(3.11) *A 3-homogeneous but not 2-transitive group preserves a linear order (isomorphic to* $(\mathbf{Q}, <)$*).*

For, since G is 2-homogeneous (by (3.2)) but not 2-transitive, a G-orbit R on ordered pairs of distinct elements satisfies trichotomy and so is (the set of directed edges of) a tournament. All of its 3-vertex sub-tournaments are isomorphic, and hence either cycles or linear orders; the former is easily seen to be impossible. Thus R is a linear order. Now some, and hence every, open interval is non-empty; so R is dense. It clearly has no end-points; so it is isomorphic to $\mathbf{Q}$.

It is also proved in Cameron (1976) that a 4-homogeneous, not 3-transitive, group is a subgroup of B or C. We saw in §2.6 that there is a k-homogeneous but not k-transitive group for every k, and also that there is a 5-homogeneous but not 4-transitive group, not contained in D. The best general result is due to Macpherson (1986b):

(3.12) *For* $k > 3$*, a* $(k+3)$*-homogeneous, not* k*-transitive permutation group is a subgroup of* D*.*

The proof of this result, like my original proof of (3.10), works mainly with finite permutation groups. (The compactness argument in the other proof breaks down irretrievably.) However, a gap still remains; closing it is likely to be very difficult.

Problem. For $k \geq 5$, is there a $(k+1)$-homogeneous and $(k-1)$-transitive but not k-transitive permutation group?

3.5. $f_n = f_{n+1}$

By (3.2), an $(n+1)$-homogeneous group is n-homogeneous. In view of the results of the last section, we may regard such groups as being reasonably well understood, and we regard the problem posed in the title of this section as settled in a particular case if we have reduced it to this situation. So we can formulate the problem thus:

What can be said about groups with $f_n = f_{n+1} > 1$?

Obviously, there is no such group with $n = 0$ (recall our convention about f_0). It is an easy exercise to show that, if this condition holds with $n = 1$, then the group has a fixed point and acts 2-homogemeously on the remaining points (so that $f_1 = f_2 = 2$), which is a satisfactory answer. We observed in §3.2 the existence of imprimitive groups with $f_2 = f_3$ (and indeed with $f_{2m} = f_{2m+1}$ for all m). Both $\mathrm{Sym}(2) \, \mathrm{Wr} \, S$ and $S \, \mathrm{Wr} \, \mathrm{Sym}(2)$ have this property, with $f_{2m} = m + 1$. Another exanple is $S \times \mathrm{Sym}(2)$, acting on the cartesian product (rather than the disjoint union) of sets of size $\aleph_0$ and 2; this group has

$$f_{2m} = f_{2m+1} = \frac{1}{2}(m+1)(m+2)$$

for all m.

The main tool available to investigate this situation is a consequence of the second (Ramsey's theorem) proof of (3.2). We associate a colour with each of the $r = f_n$ orbits on n-sets. The *colour scheme* of an $(n+1)$-set is the r-tuple whose i^{th} entry is the number of sets of the ith colour which it contains. Then just r distinct colour schemes occur (corresponding to the orbits of G on $(n+1)$-sets); and, after ordering the colours and colour schemes suitably, the colour schemes are the rows of a lower triangular $r \times r$ matrix with non-zero diagonal. (*Remark*: This is the matrix of the restriction of the linear map θ to the subspace V_n^G in the first (linear algebra) proof of (3.2). That proof gives only the less refined information that the matrix is non-singular.)

Now call the last colour "blue" and regard all other colours as shades of "red". Then we obtain a colouring with just two colours and two colour schemes, for which the corresponding 2×2 "colour scheme matrix" is again lower triangular. The colouring admits the group G; its action is transitive on blue n-sets, and on $(n+1)$-sets which contain a blue n-set.

Though the information obtained here from the linear algebra argument is less precise, it is useful in other ways, for example *via* (3.1)(ii), according to which the condition $f_n = f_{n+1}$ is preserved by passing to a supergroup.

Using these techniques, intransitive and imprimitive groups can be dealt with satisfactorily.

(3.13) *Let G be intransitive and satisfy $f_n = f_{n+1}$. Then all but at most n points lie in a single orbit of G, and G acts $(n+1)$-homogeneously on this orbit.*

(3.14) *Let G be transitive and imprimitive and satisfy $f_n = f_{n+1}$. Then n is even, and either*
(i) there is a congruence with two infinite classes, and the setwise stabiliser of these classes acts $(n+1)$-homogeneously on each; or
(ii) there is a congruence with classes of size 2, and G acts $(n+1)$-homogeneously on the set of congruence classes.

There is, however, still the possibility that something interesting happens in case (i) of (3.14): there could be a non-trivial relation between the two congruence classes which is preserved by G. Cheryl Chute Miller (to appear) has not only shown that this does indeed happen, but completely determined how it can happen (there is a unique non-trivial relation, up to isomorphism, that is, a unique closed group of this type.)

A corollary of (3.13) settles the case in which more than two consecutive f's are equal.

(3.15) *Let G satisfy $f_n = f_{n+2}$. Then the conclusions of (3.13) hold; in particular, if G is transitive, then it is $(n+2)$-homogeneous.*

This theorem gives us our first piece of information about the growth rate of the sequence (f_n): either it is ultimately constant, or it grows at least as fast as a linear function with slope $\frac{1}{2}$. The groups $S \operatorname{Wr} \operatorname{Sym}(2)$ and $\operatorname{Sym}(2) \operatorname{Wr} S$ show that this is best possible; nevertheless, it is only the first in a spectrum of growth rate results to be discussed in the next section.

We turn now to primitive groups. There are only seven known closed examples with $f_n = f_{n+1} > 1$.

Examples 1 and 2. The automorphism group of the local order L of §3.3 is 2-homogeneous and has $f_3 = f_4 = 2$. (This is shown by constructing all local orders on at most 4 points.) The same is true of its supergroup preserving or reversing the local order; this group is 2-transitive.

The next four examples are all 3-homogeneous and have $f_4 = f_5 = 2$.

Example 3. The infinite-dimensional affine group over GF(2). (A 4-set in affine space over GF(2) is either an affine plane or independent; a 5-set is either an affine plane together with one point off it, or independent.)

Examples 4 and 5. The operation of *switching* a tournament with respect to a set of points consists in reversing all instances of the relation between the set and its complement, leaving instances within or outside the set unaltered. (We saw the analogous operation for graphs in Exercise 6 of §2.10.) Example 4 is the group of switching-automorphisms of the countable homogeneous universal tournament, i.e. the permutations which map it into an image under switching (and hence fix its "switching class"). Example 5 is its supergroup in which images of the reverse tournament under switching are also allowed.

Example 6. Modify boron trees (§2.6) into "boron-carbon trees" by allowing vertices of valency 4 as well as 1 and 3. Now take the group of automorphisms of the countable homogeneous quaternary relational structure whose age consists of the structures defined on the leaves of these trees, in the same way as we did for boron trees before. (The methane molecule shows that there will be 4-sets on which no instance of the relation holds.)

Example 7. Finally, the unmodified boron trees of §2.6 give a group which is 5-homogeneous and have $f_6 = f_7 = 2$.

The strongest conjecture that could be made here is that there are no more. Weaker than this, and suggested by the examples, is that a primitive group with $f_n = f_{n+1} > 1$ is necessarily $(n-1)$-homogeneous and has $f_n = 2$. Even this appears out of reach at present. But some progress on this question has been made, and I conclude this section by surveying what is known. These results are taken from Cameron (1981), (1983b) and Cameron and Thomas (1989). The first two are specific; the third is a technical result which characterises Example 3 and is needed in the proof of the final general theorem.

(3.16) *There is no primitive group with $f_2 = f_3 > 1$.*

(3.17) *A primitive group with $f_3 = f_4 > 1$ preserves or reverses the local order L (Examples 1 and 2).*

(3.18) *Suppose that G is t-transitive and the stabiliser of t points fixes additional points. Suppose further that $f_n = f_{n+1} > 1$. Then $t = 3$, $n = 4$, and G is contained in an affine group over $\mathrm{GF}(2)$ (Example 3).*

(3.19) *Let G be primitive and satisfy $f_n = f_{n+1} > 1$. Then:*
(i) G is 2-homogeneous.
(ii) If G is t- but not $(t+1)$-homogeneous, then $n \leq 4t + 8$.
(iii) f_n is bounded by a function of n.

The bound in (3.19)(ii) has been improved by Miller (personal communication). However, it seems impossible to reduce the right-hand side below about $2t$ with present technology. The bound in (3.19)(iii) involves recursive use of Ramsey's theorem and is embarrassingly large, especially in view of the fact that the correct bound is conjectured to be 2.

Exercises

1. Describe all colourings of the points of a set Ω with r colours so that only r colour schemes of pairs occur. Deduce a strengthening of the assertion in the text, viz. a group of degree greater than 3 having $f_1 = f_2$ fixes at most one point and acts 2-homogeneously on its non-fixed points.

2. Show that, if $|\Omega| > 5$, then a colouring of the 2-subsets of Ω with two colours and two colour schemes of 3-sets is one of the following, up to choice of colours:
 (a) there is a partition of Ω into two parts so that edges within each part are red, and those between the parts are blue;
 (b) blue edges are pairwise disjoint.
Deduce (3.16), and also the case $n = 2$ of (3.14).

Remark: All colourings of 2-sets with r colours, having just r colour schemes of 3-sets, are described in Cameron (1982), which can be used as a crib for this exercise.

3. (i) Show that, if $a, b > n$ and $G = \mathrm{Sym}(a) \times \mathrm{Sym}(b)$ (in its intransitive representation), then $f_{n+1}(G) > f_n(G)$.
 (ii) Show that, if $a, b > 2$ and $G = \mathrm{Sym}(a) \,\mathrm{Wr}\, \mathrm{Sym}(b)$ (in its imprimitive representation), then $f_{n+1}(G) > f_n(G)$.
 (iii) Using these and (3.1)(ii), deduce part of (3.13) and (3.14).

Remark: The argument can be completed by using, in place of the symmetric groups, the groups induced on the relevant sets by G.

4. (Concerning Examples 1 and 4.) A tournament on three points can be switched into just one of the two possible 3-cycles. So, with any tournament is associated a ternary relation giving the sense of the 3-cycle on each of its triples. Call this ternary relational structure the *T-structure* associated with the tournament. Prove the following.

(i) A tournament is a local order if and only if its T-structure is a circular order.

(ii) Two tournaments are switching-equivalent if and only if their T-structures are equal.

(iii) The class of finite T-structures has the amalgamation property.

(iv) The countable universal homogeneous T-structure is the T-structure of the countable universal homogeneous tournament. (This is the structure whose automorphism group is Example 4.)

3.6. GROWTH RATES

The picture presented in §§3.2, 3.3 shows a spectrum of possible rates of growth of realised sequences (f_n). with gaps between them. Though our knowledge is far from complete, we do know that this is not accidental, and that some at least of the gaps really do occur. These results are due principally to Macpherson (1985a), (1985b), (1987), with some contributions from others such as Pouzet (1981).

The first result of this kind was mentioned after (3.16): either (f_n) is ultimately constant, or it is bounded below by a linear function with slope $\frac{1}{2}$. But this is only the first step in an infinite sequence:

(3.20) *Either*
(i) $c_1 n^d \le f_n \le c_2 n^d$ for all n, where $d \in \mathbb{N}$ and $c_1 > 0$; or
(ii) for any $\epsilon > 0$, $f_n > \exp(n^{\frac{1}{2}-\epsilon})$ for all sufficiently large n.

In other words, there is an ω-sequence of possible growth rate bands (polynomial with non-negative integer degree) followed by a gap to fractional exponential (comparable with the partition function). It is reasonable to conjecture that, in (i), it is true that for some $c \in \mathbf{R}$, $c > 0$, and $d \in \mathbf{N}$,

$$\lim_{n \to \infty} f_n / cn^d = 1,$$

but this has not been proved. If true, then a further question would be: what are the possible values of c, for given d?

For primitive groups, the gap is even more striking:

(3.21) *If G is primitive, then either G is highly homogeneous, or $f_n > c^n$ for all sufficiently large n, where c is an absolute constant.*

Macpherson gives $c = 2^{\frac{1}{5}} - \epsilon$, and speculates that improvement is possible with more care. The best known examples (see Example 1 in §3.3) have growth just less than 2^n.

There is some slight evidence of another ω-sequence of growth rates below exponential for imprimitive groups: the possibilities would be

$$\exp(n^{(d+1)/(d+2)-\epsilon}) < f_n < \exp(n^{(d+1)/(d+2)+\epsilon})$$

for $d \in \mathbf{N}$. We saw in §3.2 that all of these values are realised. Moreover, there are infinitely many sequences with growth rate faster than fractional exponential but slower than exponential.

For growth rates beyond exponential, there is evidence of further gaps, at least under additional hypotheses. Let M be $\aleph_0$-categorical. M is said to have the *independence property* if there is a relation $R(\bar{x}, \bar{y})$ definable in M (that is, a union of orbits of $\mathrm{Aut}(M)$), and two sets A and B of tuples of the appropriate arities, such that A is indexed by $\mathbf{N}$, B by the set of finite subsets of $\mathbf{N}$, and for $\bar{a} \in A$, $\bar{b} \in B$, the relation $R(\bar{a}, \bar{b})$ holds if and only if the index of $\bar{a}$ is a member of the index of $\bar{b}$. This strange-looking hypothesis, somewhat similar to instability, was introduced by Shelah in the course of his investigations into the numbers of models of a theory. Its precise significance in the context of permutation group theory is not clear to me, but the next result indicates that it has some effect!

(3.22) *(i) If M is homogeneous over a finite language and*

$$f_n(\mathrm{Aut}(M)) > 2^{n^{1+\epsilon}}$$

for sufficiently large n (where $\epsilon > 0$), then M has the independence property.
(ii) If M is $\aleph_0$-categorical and has the independence property, then

$$f_n(\mathrm{Aut}(M)) > 2^{P(n)}$$

for all n, where P is a polynomial of degree 2.

Of course, the possibility of growth rates between $2^{n^{1+\epsilon}}$ and 2^{cn^2} is left open; but a structure with such growth could not be homogeneous over a finite language.

The next result requires the definitions of ω-stability and strictly minimal sets, which are perhaps even less enlightening than that of the independence property at first sight. These concepts will be discussed a little further in §5.3. For now, note merely that $\aleph_0$-categorical and ω-stable structures involve, and are in a sense constructed around, certain strictly minimal sets which are either disintegrated (infinite sets without structure, with the symmetric group acting), or projective or affine spaces over finite fields (with the appropriate groups). We saw in §3.3 that, for such groups over $\mathrm{GF}(q)$, the growth of f_n is roughly like $q^{\frac{1}{4}n^2}$.

(3.23) *Let M be $\aleph_0$-categorical and ω-stable, and $\mathrm{Aut}(M)$ not highly transitive. Then $f_n > \lfloor n/5 \rfloor!$ for all n. Moreover, either*
(i) $f_n > 2^{P(n)}$ for all n, where P is a polynomial of degree 2; or
(ii) $f_n < 2^{n^{1+\epsilon}}$ for large n.

Perhaps, in view of the origin of these concepts in the context of the enumeration of infinite models of a theory, it is not totally unexpected that they are relevant to counting finite substructures of a structure.

What about upper bounds? The next result shows that general growth rates are unbounded, but that for homogeneous structures over finite languages we have already seen the worst.

(3.24) *(i) For any sequence (a_n), there is an oligomorphic group G for which $f_n(G) > a_n$ for all $n > 0$.*
(ii) If $G = \mathrm{Aut}(M)$, where M is homogeneous over a finite language, then $f_n(G) \leq 2^{P(n)}$, where P is a polynomial.

For (i), take a language with a_n relations of arity n for every n, and let C be the class of finite structures over this language (with no restriction on the interpretation except that a relation holds only if its arguments are all distinct). Then C satisfies Fraïssé's conditions, and has rather more than a_n n-element structures for all n.

To prove (ii), observe that, if the relations in the language have arities $k_1, \ldots, k_r$, then the number of n-element structures over the language does not exceed

$$2^{n^{k_1} + \cdots + n^{k_r}}.$$

This leaves open the question of a possible converse to (3.24)(ii). Consider the projective space over GF(2), for which f_n is very roughly $2^{\frac{1}{4}n^2}$. The points can be identified with the non-zero vectors in a vector space over GF(2) of countable dimension. I claim that this structure is not homogeneous over any finite language. For, given n, let $\bar{a}$ be a linearly independent n-tuple, and $\bar{b}$ an n-tuple with sum 0 but satisfying no other linear dependence relation. Then $\bar{a}$ and $\bar{b}$ lie in different orbits, but cannot be distinguished by any relation of arity less than n. So, to make the structure homogeneous, we need a relation of arity at least n for every natural number n.

It may be, however, that if $f_n(G)$ grows sufficiently slowly (at most exponentially, perhaps), and G is closed, then G is the automorphism group of a homogeneous structure over a finite language.

I will try to give a very brief indication of the techniques of proof used by Macpherson. I won't discuss Pouzet's arguments with sub-ages which establish the polynomial growth rates in (3.20). Macpherson takes over once the growth is faster than polynomial.

At this point, and important tool becomes available, namely Exercise 2 in §3.3: we can describe growth rates with a sufficiently broad brush that a group has "the same" growth as the stabiliser of a point. (That is, sequences whose growth differs by only a polynomial factor are identified.) Consider (3.21). Suppose first that, for some integer $s \geq 4$, we have managed to show that any primitive group with growth rate slower than $c^n/P(n)$ (where c is a fixed constant and P a polynomial) is s-homogeneous. Let G be such a group. If G is not 3-transitive, then (by (3.12) and the following remark) it preserves a betweenness relation or a circular order, and it is easy to show that either G is highly homogeneous or its growth is faster than $2^n/P(n)$. We will assume that $c < 2$, so that the second alternative is impossible; and if the first holds, then we have finished. On the other hand, if G is 3-transitive, then the stabiliser of a point is 2-transitive, and hence primitive; and it has slow growth, so by assumption it is s-homogeneous and G is $(s+1)$-homogeneous. Thus, we only have to get as far as 4-homogeneity.

The 4 can be reduced to 3 by making use of the following result of Cameron (1983b) (proved for just this purpose):

(3.25) *Suppose that G is 3-homogeneous and 2-transitive, and that the stabiliser of a point in G is imprimitive on the remaining points. Then either G preserves a betweenness relation, or G is contained in the point stabiliser of the "boron trees" group of §2.6.*

In the latter case, as we saw, the growth rate is about $(2.483\ldots)^n$.

This leaves the problem of showing that primitivity and slow growth together imply 3-homogeneity; this is where the real work is needed. If G is not 2-homogeneous, then it leaves invariant a graph (whose edges can be taken to be any G-orbit on 2-sets); similarly, if G is not 3-homogeneous, it leaves invariant a 3-hypergraph. Thus Macpherson has to consider primitive groups of automorphisms of graphs and hypergraphs. I shall only discuss graphs; for 3-hypergraphs, the arguments are somewhat similar but even more complicated.

First, it is shown that the neighbour sets of any two vertices have infinite symmetric difference. (This is a general fact about infinite graphs admitting primitive groups with $f_n < \infty$.) Hence, given any finite set X of vertices, it is possible to find infinite sets $T_1, T_2, \ldots$ such that two vertices in the same set T_i have the same neighbours within X, while two vertices in X are joined to different T_i. Then we use Ramsey's theorem to make each T_i complete or null, and to "standardise" the edges between them in a suitable way. This done, it is possible to "encode" finite objects (such as partitions or trees) by subgraphs whose size is not too much larger than that of the objects being encoded, so that many non-isomorphic subgraphs are produced. The argument is quite a bit more complex than I've indicated.

It can also be see, in general terms, how hypotheses like instability or the independence property become involved — these hypotheses assert that certain objects can be encoded into our structure in the right way. For example, the bipartite form of Ramsey's theorem asserts that an infinite bipartite graph contains a subgraph which is null, or complete bipartite, or isomorphic to the graph with vertex set $\{v_i, w_i : i \in \mathbf{N}\}$ in which v_i is joined to w_j if and only if $i < j$; and in the third case, the graph is unstable. (The terms "infinite bipartite graph", "infinite null graph", etc. are meant to imply that both bipartite blocks are infinite.)

Exercises

1. Prove the assertion in the preceding paragraph.

2. Let G be a subgroup of the automorphism group of an infinite linear order. Show that either G is highly homogeneous, or $f_n(G) \geq 2^n/P(n)$ for some polynomial P. [*Hint*: Reduce to the case where G is intransitive.] Deduce that the same assertion holds for a group of automorphisms of a circular order.

3. Find a set of relations which "homogenize" the projective space over GF(2). (See the remarks after (3.24).)

3.7. APPENDIX: CYCLE INDEX

In this appendix, I explain, among other things, how the sequence (f_n) for a wreath product is calculated from information about the factors.

The Redfield-Pólya theory of combinatorial enumeration is based on the cycle index of a finite permutation group G. This is a polynomial in variables $s_1, \ldots, s_n$ (where $n = |\Omega|$), defined by

$$Z(G; s_1, \ldots, s_n) = \frac{1}{|G|} \sum_{g \in G} s_1^{c_1(g)} \ldots s_n^{c_n(g)},$$

where $c_i(g)$ is the number of cycles of length i in the cycle decomposition of the permutation g. Note that, if the definition is extended to (not necessarily faithful) group actions, the result only depends on the permutation group which is induced (the kernel of the action).

Its important property is that many problems involving enumeration of group orbits are solved by specialising the cycle index. Suppose that we have a collection F if "figures" with non-negative integer weights, there being a_n figures of weight n. The *figure-counting series* is the generating function

$$a(t) = \sum_{n \geq 0} a_n t^n.$$

We consider functions f from Ω to the set of figures. Each function has a weight (the sum of the weights of its values), and there is an action of G on the set of functions, defined by

$$fg(\alpha) = f(\alpha g^{-1}).$$

If b_n is the number of G-orbits on the set of functions of weight n, then the *function-counting series* is the generating function

$$b(t) = \sum_{n \geq 0} b_n t^n.$$

(3.26) $b(t) = Z(G; a(t), \ldots a(t^n))$.

This is a standard application of the orbit-counting lemma (often, incorrectly, called "Burnside's Lemma"). A typical application runs as follows. Let G be the symmetric group of degree n, in its action on the set Ω of 2-element subsets of $\{0, \ldots, n-1\}$. Take a set of two figures, called "edge" (weight 1) and "non-edge" (weight 0). The figure-counting series is thus $1 + t$; so the function-counting series, namely

$$Z(G; 1 + t, 1 + t^2, 1 + t^3, \ldots,)$$

is the generating function for the number of graphs with a fixed number n of vertices and a variable number of edges, up to isomorphism. The required cycle index can be written down explicitly, as a sum over partitions of n, with a little care.

Note that Z is the unique polynomial for which the result of (3.26) holds.

Now let G be an oligomorphic permutation group, or a permutation group of finite degree. We define the *modified cycle index* $\tilde{Z}(G)$ of G as follows. List the (finite) permutation groups induced on finite subsets of Ω by their set-wise stabilisers in G, one for each orbit of G on finite sets; and sum their cycle indices. We obtain a formal power series in the indeterminates $s_1, s_2, \ldots$. (Note that it is well-defined; for a given finite product of indeterminates occurs only in cycle indices of groups of one possible degree, namely its "weight" — obtained by summing the indices of those indeterminates s_i occurring in the product with the correct multiplicities — and by the oligomorphy assumption, there are only finitely many terms containing this product. By convention, the empty set contributes 1 to the sum.)

If G is the automorphism group of a homogeneous structure M (necessarily $\aleph_0$-categorical), then the modified cycle index of G depends only on the age of M. In fact, it is obtained by summing the ordinary cycle indices of the automorphism groups of all the unlabelled structures in $\mathrm{Age}(M)$. (In this form, it would be possible to extend the definition to a wider class of structures, but I shall not do so; however, see Joyal (1981) for this development and much more.)

For finite permutation groups, we obtain nothing new, although the relationship is not obvious:

(3.27) *If G is finite, then*

$$\tilde{Z}(G;\, s_1,\, s_2. \ldots) = Z(G;\, s_1 + 1,\, s_2 + 1,\, \ldots).$$

A formal proof is not hard, but the result can be seen as follows. Given any set F of figures, add a new distinguished figure $*$ of weight zero. The function-counting series for the new set is obtained by substituting $a(t)$, $a(t^2)$, $\ldots$ in the right-hand side of (3.27), where $a(t)$ is the figure-counting series for F. But it is also obtained by the same substitution in the left-hand side, as can be seen by considering, for each function, the set of points on which its value is not $*$. The result now follows from the uniqueness of the cycle index.

We are interested in this series mainly for infinite permutation groups. Two special-isations are particularly important:

(3.28) *(i)* $f_G(t) = \tilde{Z}(G; t, t^2, t^3 \ldots)$.
(ii) $F_G(t) = \tilde{Z}(G; t, 0, 0, \ldots)$.

The first holds because $Z(G; t, t^2, \ldots) = t^n$ for a finite permutation group of degree n — take a single figure of weight 1. The second follows from the facts that

$$Z(G; t, 0, \ldots) = t^n/|G|,$$

and that $n!/|G|$ is the number of G-orbits on orderings of its domain, for a finite group G of degree n.

Now modified cycle indices of direct and wreath products and of stabilisers can be computed as follows:

(3.29) *(i)* $\tilde{Z}(G \times H) = \tilde{Z}(G)\tilde{Z}(H)$.
(ii) $\tilde{Z}(G \operatorname{Wr} H) = \tilde{Z}(H; \tilde{Z}(G) - 1)$.
(iii) If G is transitive then $\tilde{Z}(G_\alpha) = \frac{\partial}{\partial s_1}\tilde{Z}(G)$.

In (ii), the substitution $A(B)$ of a formal power series into another is defined to mean

$$A(B(s_1, s_2, \ldots), \ B(s_2, s_4, \ldots), \ B(s_3, s_6, \ldots) \ldots).$$

This is well-defined provided that the constant term of B is zero.

In (iii), if G is not transitive, then the derivative on the right gives the sum of the modified cycle indices of the stabilisers of points, one from each G-orbit.

As a corollary, we obtain results (3.5) and (3.6). We observed then that the sequence $(f_n(G \operatorname{Wr} H))$ is not determined by $(f_n(G))$ and $(f_n(H))$. The appropriate formula is given by the next result, which follows immediately from (3.28)(i) and (3.29)(ii):

(3.30) $f_{G \operatorname{Wr} H}(t) = \tilde{Z}(H; f_G(t) - 1, f_G(t^2) - 1, \ldots,)$.

So we need to know $(f_n(G))$ and the modified cycle index of H.

The modified cycle indices of $S = \operatorname{Sym}(\Omega)$ and $A = \operatorname{Aut}(\mathbf{Q}, <)$ are given by the formulae:

(3.31) *(i)* $\tilde{Z}(S) = \exp(-\sum_{n \geq 1} s_n/n)$.
(ii) $\tilde{Z}(A) = 1/(1-s_1)$.

Proof. For (i), we start from the fact that Sym(n) contains

$$n! / \prod_{i \geq 1} i^{c_i} c_i!$$

permutations with c_i cycles of length i, where $\sum_{i \geq 0} ic_i = n$. We must multiply this expression by $\prod_{i \geq 1} s_i^{c_i}$, divide by $n!$, and sum over all sequences (c_i) satisfying $\sum_{i \geq 1} ic_i = n$, and then sum over n, to obtain the required modified cycle index. The result is

$$\sum_{(c_i)} \prod_{i \geq 1} \frac{(s_i/i)^{c_i}}{c_i!},$$

where the summation is over all sequences (c_i) with all but finitely many terms zero. This can be rearranged, using the distributive law, to

$$\prod_{i \geq 1} \sum_{c \geq 0} \frac{(s_i/i)^c}{c!}$$

$$= \prod_{i \geq 1} \exp(\frac{s_i}{i})$$

$$= \exp \sum_{i \geq 1} (\frac{s_i}{i}).$$

(ii) is straightforward: the cycle index of the trivial group of degree n is s_1^n.

(3.7) follows from this result and (3.30). Part (ii) is immediate, while (i) requires a little further calculation, as follows.

$$f_{G \text{ Wr } S}(t) = \exp \sum_{i \geq 1} \frac{(f_G(t^i) - 1)}{i}$$

$$= \exp \sum_{i \geq 1} \sum_{j \geq 1} \frac{f_j(G) t^{ij}}{i}$$

$$= \exp \sum_{j \geq 1} f_j(G) \sum_{i \geq 1} \frac{t^{ij}}{i}$$

$$= \exp \sum_{j \geq 1} -f_j(G) \log(1 - t^j)$$

$$= \prod_{j \geq 1} (1 - t^j)^{-f_j(G)}.$$

Here is a curious application of (3.28). Let C denote the group of permutations of the roots of unity which preserve the cyclic order (one of the highly homogeneous groups of (3.10)). The group induced on an n-set by its setwise stabiliser in C is C_n, the cyclic group of order n. Thus, we have

$$\check{Z}(C) = 1 + \sum_{n \geq 1} Z(C_n)$$

$$= 1 + \sum_{n \geq 1} \sum_{d|n} \phi(d) s_d^{n/d}/n$$

$$= 1 - \sum_{d \geq 1} \frac{\phi(d)}{d} \log(1 - s_d),$$

where ϕ is Euler's totient.

Now $\check{Z}(C; t, t^2, \ldots,) = f_C(t) = 1 + t/(1 - t)$; so

$$\prod_{d \geq 1} (1 - t^d)^{-\phi(d)/d} = \exp(t/(1 - t)),$$

a well-known identity which is closely related to the similar-looking identity

$$\prod_{d \geq 1} (1 - t^d)^{-\mu(d)/d} = \exp(t),$$

where μ is the Möbius function.

Before leaving the subject, I shall briefly describe how the exponential constant 2.483... for boron trees can be calculated. As described in §3.3, the constant is unaltered if we use instead the stabiliser of a point, for which the f_n are the Wedderburn-Etherington numbers: f_n is the number of binary trees with left and right not distinguished. Such a tree with more than one leaf has two interchangeable subtrees of the same form above its root. So, taking $a(t) = f_G(t) - 1$ as figure-counting series and group Sym(2), the function-counting series is just $a(t) - t$. Thus we have

$$a(t) = t + Z(\text{Sym}(2); a(t), a(t^2))$$

$$= t + \frac{1}{2}(a(t)^2 + a(t^2)),$$

as claimed earlier.

Now the exponential constant is the reciprocal of the radius of convergence of $a(t)$. The radius of convergence is the distance from the origin to the nearest singularity. Now we have

$$a(t)^2 - 2a(t) + (a(t^2) + 2t) = 0,$$

so

$$a(t) = 1 \pm \sqrt{1 - 2t - a(t^2)}.$$

Since $a(0) = 0$, we must take the minus sign here. We see that the relevant singularity is a branchpoint, occurring at $t = r$, where $a(r^2) = 1 - 2r$. Since $r < 1$, we have $r^2 < r$, and so $a(r^2)$ is well-behaved and can be approximated by a Taylor polynomial. (The coefficients f_n of this polynomial can be calculated from the recurrence relation

$$f_n = f_1 f_{n-1} + f_2 f_{n-2} + \dots,$$

where the last term is

$$\begin{cases} f_{(n-1)/2} f_{(n+1)/2}, & n \text{ odd}, \\ \binom{f_{n/2}+1}{2}, & n \text{ even}, \end{cases}$$

as explained in §3.4.) Then the equation $a(r^2) = 1 - 2r$ can be solved numerically.

It would take too long a detour to discuss how more accurate asymptotic expansions are calculated; I refer to Comtet (1974) or Harary and Palmer (1973).

3.8. APPENDIX: A GRADED ALGEBRA

This construction is a natural outgrowth of the first (linear algebra) proof of (3.2).

Let Ω be an infinite set. As in §3.1, let V_n be the **Q**-vector space of functions from the set of n-subsets of Ω to **Q**. Let $A = V_0 \oplus V_1 \oplus \dots$. Now we make A into an algebra by defining multiplication on the homogeneous components by the rule

$$fg(K) = \sum \{f(A)g(K \setminus A) : A \subseteq K, |A| = n\}$$

for $f \in V_n$, $g \in V_m$, $|K| = n + m$ (so that $fg \in V_{n+m}$).
Extending this multiplication linearly to the whole of A, we obtain a graded algebra, which is easily seen to be commutative and associative. (It is a "reduced incidence algebra", in the sense of Rota (1964).)

The algebra A has a 1, namely the function in V_0 with value 1. Let e denote the function in V_1 which takes the constant value 1. Then multiplication by e induces the linear map θ in the proof of (3.2), from V_n to V_{n+1}. Since θ is a monomorphism, e is a non-zero-divisor.

The algebra A has many zero-divisors; indeed, many nilpotent elements. Any function f with finite support is nilpotent: $f^n = 0$ if the support of f doesn't contain n disjoint sets.

Now let a group G act on Ω. Then G acts on the algebra A (by permuting the arguments of the functions, viz. $(fg)(K) = f(Kg^{-1})$). Let

$$A^G = V_0^G \oplus V_1^G \oplus \dots$$

be its set of fixed points. Then A^G is a graded algebra containing 1 and e; we have $\dim(V_n^G) = f_n(G)$, if this number is finite.

If G has a finite orbit, then A^G obviously contains nilpotent elements.

Conjecture: If G has no finite orbits, then e is prime in A^G (and hence A^G is an integral domain).

No progress has been made on this conjecture since it was made. An application of its truth, however, is given by Cameron (1981): it is shown, modulo the conjecture, that, if $f_n(G) = f_{n+1}(G) > 1$, and n is minimal subject to this, and if G is transitive, then G is $\lceil \frac{1}{2}n \rceil$-homogeneous. (This would strengthen (3.19)(ii).)

In some special cases, the structure of A^G can be determined. There are two situations in which this is the case. The first is described in the next result.

(3.32) *(i)* $A^{G \times H} = A^G \otimes A^H$.

(ii) If S is the symmetric group, then A^S is the polynomial ring in one indeterminate e.

(iii) If G is the direct product of r copies of S, then A^G is a polynomial ring in r homogeneous generators of degree 1.

(iv) If $G = S \operatorname{Wr} H$, where H is a finite permutation group, then A^G is the ring of invariants of H (regarded as a linear group, using the representation by permutation matrices.)

For example, if $G = S \operatorname{Wr} \operatorname{Sym}(r)$, then A^G is the ring of symmetric functions in r indeterminates, and is a polynomial ring in homogeneous generators of degrees 1, 2, $\ldots, r$.

Another case concerns certain homogeneous structures. We require that, in the age of the structure, there is a notion of "connectedness", so that any structure is a "disjoint union" of "connected components"; and also a notion of "involvement" (corresponding to the concept of spanning subgraph in graph theory), so that, given a structure X and any partition of its domain, M "involves" the disjoint union of the induced substructures on the parts of the partition. (This can be given a precise axiomatic formulation.) The conditions certainly hold for the uniform hypergraphs R_k, and for Henson's graphs G_n (see §4.4). Under these hypotheses, it can be shown that A^G is a polynomial ring; its homogeneous generators are the characteristic functions of the "connected" structures in the age.

An important special case is the following.

(3.33) *For any oligomorphic permutation group G, it holds that $A^{G \operatorname{Wr} S}$ is a polynomial ring with $f_n(G)$ homogeneous generators of degree n for all $n \geq 1$.*

In this case, an object in the age of the canonical $(G \operatorname{Wr} S)$-structure has a natural partition (corresponding to the congruence for $G \operatorname{Wr} S$); an object is connected if and only if it has just a single part (i.e. it is contained in a single class of the congruence), in which case it corresponds to an object in the age of the canonical A^G-structure. Note that the algebraic structure of $A^{G \operatorname{Wr} S}$ depends on A^G only in a crude numerical way.

Note that, if A^G is a polynomial ring in a finite number r of generators, then $f_n(G)$ grows like a polynomial of degree $r - 1$. The converse is not true, since rings of invariants are not usually polynomial rings. Is there any kind of partial converse?

Exercise

Prove the assertion implicit in the Conjecture, viz. if e is a prime element in A^G, then A^G is an integral domain.

4

Subgroups

4. Subgroups

4.1. BEGINNINGS

This chapter is about subgroups of automorphism groups of various structures.

There are several aspects to note. For a start, as we saw in Chapter 2, if a group G is the automorphism group of a structure, then it is the automorphism group of a homogeneous structure; and this is equivalent to G being a closed subgroup of the symmetric group. Describing all subgroups of the symmetric group is too wide a task, so I'll restrict both the structures, and the kinds of subgroups considered. As to the first, I shall consider only

(a) $\aleph_0$-categorical structures (those whose automorphism groups are oligomorphic); and

(b) homogeneous structures whose age has the strong amalgamation property (those for which the stabiliser of a tuple in the automorphism group fixes no additional points).

Similar results hold in other cases; a notable example of this is provided by recursively saturated structures. (The results for these structures, due to Richard Kaye, were obtained following the Durham symposium.)

The results will in the main be constructions of subgroups with various properties, but at the end of the chapter I will describe some restrictive results characterising certain kinds of subgroups (normal subgroups, subgroups of small index, etc.).

There are differing levels of detail about a subgroup. We could be concerned with its structure as abstract group, as permutation group on some subset of the domain, or as permutation group on the entire domain. I shall talk mainly about the third; but I shall close this introductory section with sample results about the first two.

Many techniques are used in the constructions. These include direct construction of automorphisms "a point at a time", the use of measure and category, and expansion of the language (especially in cases where the strong amalgamation property holds). In successive sections I discuss these methods and give some results proved with their use.

First, however, the promised "classical" subgroup theorem.

(4.1) *Let M be $\aleph_0$-categorical. Then $\mathrm{Aut}(M)$ contains a subgroup isomorphic to $A = \mathrm{Aut}(\mathbf{Q}, <)$, acting on a subset of M in the way that A acts on $\mathbf{Q}$.*

The proof comes easily from the Ehrenfeucht-Mostowski theorem (1956), according to which, if Σ is a theory possessing infinite models and Λ an ordered set, then there is a model M of Σ containing Λ such that any order-preserving permutation of Λ can be extended to an automorphism of M. The proof actually shows that M is generated by Λ (over an extended language containing "Skolem functions"). From this, we further conclude that M is countable if Λ and the language of Σ are countable, and that $\mathrm{Aut}(\Lambda)$ acts faithfully on the copy of Λ, so that it is embedded as a subgroup of $\mathrm{Aut}(M)$, rather than as a quotient of a subgroup. Now we take $\Lambda = \mathbf{Q}$ and Σ the theory of our given $\aleph_0$-categorical structure; then the countable M given by the Ehrenfeucht-Mostowski theorem is isomorphic to the given structure, and its automorphism group embeds A in the manner required.

(4.2) *If M is $\aleph_0$-categorical, then $\mathrm{Aut}(M)$ contains a free group of rank $2^{\aleph_0}$.*

By (4.1), it suffices to prove this for the $\aleph_0$-categorical structure $(\mathbf{Q}, <)$. I learnt the following argument from Jim Kister.

We need the result (due to Sierpiński) that a countable set contains $2^{\aleph_0}$ infinite, almost disjoint subsets. (Two sets are *almost disjoint* if their intersection is finite.) For the simplest proof, take the countable set $\mathbf{Q}$, and for each irrational real number α, take the set S_α of terms in some Cauchy sequence of rationals converging to α. (Sequences with different limits can have only finitely many terms in common.)

We also need the fact that $\mathrm{Aut}(\mathbf{Q}, <)$ contains a free subgroup of countable rank. It suffices to find a free subgroup of rank 2, since the derived subgroup of such a group is free of countable rank. There are several proofs of this fact. We will see some in the course of this chapter. Alternatively, we can use the recent result that the order-preserving permutations $x \mapsto x + 1$ and $x \mapsto x^3$ of $\mathbf{R}$ generate a free group of rank 2 (White (1988)); the argument doesn't work directly for $\mathbf{Q}$, but the existence of two order-preserving permutations satisfying no non-trivial relation is first-order, and so can be transferred down to $\mathbf{Q}$ using the Löwenheim-Skolem theorem (1.6).

Yet again, it is well-known that a free group admits a linear order invariant under right multiplication; and it is easy to check that this order is dense (see Neumann (1949)).

Now we embark on the proof of (4.2). Let $\{S_\alpha : \alpha < 2^\omega\}$ be an almost-disjoint family of subsets of $\mathbf{N}$. (We are, as usual, regarding 2^ω, the least ordinal of cardinal $2^{\aleph_0}$, as the set of all smaller ordinals.) Let s_α be the function from $\mathbf{N}$ to $\mathbf{N}$ which enumerates the elements of S_α in increasing order. For each $n \in \mathbf{N}$, let $f_{n0}, f_{n1}, \ldots$ be generators of a free group of order-preserving permutations of the rational interval $(n, n + 1)$. (This interval is order-isomorphic to $\mathbf{Q}$.) For $\alpha < 2^\omega$, let g_α be the permutation of $\mathbf{Q}$ which acts on $(n, n+1)$ as $f_{n s_\alpha(n)}$ for all $n \in \mathbf{N}$. Then the $2^{\aleph_0}$ permutations g_α generate a free group. For take a non-trivial word in $\alpha_1, \ldots, \alpha_r$. By almost-disjointness, there exists $n \in \mathbf{N}$ such that $s_{\alpha_1}(n), \ldots, s_{\alpha_r}(n)$ are all distinct. Then the given word acts non-trivially on $(n, n + 1)$.

Exercises

1. Find two order-automorphisms of $\mathbf{Q}$ which generate a free group.

2. Show that the cartesian product of countably many free groups of countable rank contains a free group of rank $2^{\aleph_0}$.

4.2. A THEOREM OF MACPHERSON

Bill Kantor pointed out to me some years ago that the free group of countable rank has a faithful highly transitive permutation representation. (I'll give the simple proof of this later.) Of course, a group is highly transitive if and only if it is dense in the symmetric group. Noting that the symmetric group is the automorphism group of a set without structure, the obvious generalisation is (4.3) below, which was proved by Dugald Macpherson.

(4.3)   *The automorphism group of an $\aleph_0$-categorical structure contains a dense subgroup which is free of countable rank.*

I won't give Macpherson's proof of this; it is somewhat bare-handed, and I am convinced that a more conceptual proof should be possible. (More about failed attempts to find such a proof later.)

If M is homogeneous, then a dense subgroup of $\text{Aut}(M)$ is one which acts "homogeneously" on M; that is, any isomorphism between finite subsets of M can be extended to an element of G.

In some cases, it is possible to find a dense subgroup which is free of rank 2. This seems to require much more intricate argument when it is possible. This assertion was proved for the symmetric group by McDonough (1977), for the universal hypergraphs R_k by Macpherson (1986b), and for $\text{Aut}(\mathbf{Q}, <)$ in a very recent preprint by Glass and McCleary (see also McCleary (1985)).

Problem. Find a necessary and sufficient condition on the $\aleph_0$-categorical structure M for $\text{Aut}(M)$ to have a dense free subgroup of rank 2.

Macpherson points out that not all $\aleph_0$-categorical structures have this property. Let M be the disjoint union of a finite structure X with automorphism group H, and an infinite set without further structure. A dense subgroup of $\text{Aut}(M)$ induces H on X, and so requires at least as many generators as H does. But H may be any finite group.

Obviously an infinite cyclic group cannot act densely on M in any but trivial cases. But which permutation representations of cyclic groups can occur as subgroups of $\text{Aut}(M)$? In other words, what are the possible cycle structures of automorphisms of M? This question has been answered in some cases, notably for the random graph R (and some of its relatives) by John Truss (1989b). I'll return to this matter later.

Exercises

1. Construct an $\aleph_0$-categorical structure whose automorphism group contains no finitely generated dense subgroup.

2. (i) Show that non-identity permutations in $A = \text{Aut}(\mathbf{Q}, <)$ have infinitely many infinite cycles and no finite cycles other than fixed points, but the number of fixed points is arbitrary (finite or infinite).

   (ii) Find the cycle structures of elements in the other four highly homogeneous groups of the conclusion to (3.10).

4.3. THE RANDOM GRAPH REVISITED

In 1963, Erdös and Rényi proved the following paradoxical result. Throughout this section, R will denote the unique countable homogeneous graph whose age consists of all finite graphs. (This is the graph considered at some length in §2.10.)

By a *random* countable graph, I mean one which is chosen by the following procedure: take a countably infinite set of vertices (which might as well be **N**); list all unordered pairs of vertices in an ω-sequence; and then choose, independently with probability $\frac{1}{2}$ for each pair in the list, whether the vertices in the pair are joined by an edge or not. Somewhat more pedantically, it is a random variable on the standard probability space of ω-sequences of zeros and ones (as explained in §1.4), regarding the vertex set and the enumeration of its pairs as fixed.

(4.4) *With probability 1, a countable random graph is isomorphic to R.*

In other words, there is only one countable random graph! Erdős and Spencer describe this result as demolishing the theory of countable random graphs.

This result is basic to what follows, so I'll give the proof. (It is my contention that mathematics is unique among academic pursuits in that such an apparently outrageous claim can be made completely convincing by a short argument.) In fact, half the work is already done. We know that R is characterised up to isomorphism (as countable graph) by condition (♣) of §2.10 (repeated below); so all we have to do is to show that (♣) holds with probability 1.

(♣) *Given any finite subset A of the vertex set, and any subset B of A, there exists a vertex whose neighbour set in A is precisely B.*

In other words, we have to show that the event that (♣) fails has probability 0. Now there are only countably many choices for the pair (A, B) of finite sets; and the union of countably many null sets is a null set. So it suffices to prove that the failure of (♣) for a *particular* pair (A, B) is null.

Given A and B, with $|A| = m$, the probability that a vertex z outside A does *not* satisfy the conclusion is $1 - 1/2^m$. Since these events are independent for different choices of z, and there are infinitely many vertices available, the probability that no vertex has the correct joins is

$$\lim_{n \to \infty} \left(1 - \frac{1}{2^m}\right)^n = 0.$$

There is a suggestive way of expressing this result. Our description of the random graph amounts to a probability measure on the space $\mathcal{G}$ of all graphs on a given countable vertex set **N**. Now the symmetric group Sym(**N**) acts on $\mathcal{G}$; its orbits are the isomorphism types, and (4.4) asserts that there is a single orbit whose complement is a null set. We could say that Sym(**N**) is "almost transitive" on $\mathcal{G}$. Furthermore, for every n, there is a product measure on $\mathcal{G}^n$. An element of $\mathcal{G}^n$ is an n-tuple of graphs

on **N**, and so is a structure over a language with n binary relations (interpreted as
graphs, that is, irreflexive symmetric relations). There is a countable homogeneous
object $R(n)$ in $\mathcal{G}^n$ whose age consists of all n-tuples of finite graphs (the conjunction
of n copies of the age of R, in the notation of (3.9)). It is easily shown that each of
the n graphs in $R(n)$ is itself homogeneous and universal, and hence isomorphic to
R. Once again, the orbit of $R(n)$ has measure 1 in $\mathcal{G}^n$. So we could say that Sym(**N**)
is "almost n-transitive" on $\mathcal{G}$ for all n, in other words, is "almost highly transitive".
As far as I know, no theory of almost highly transitive groups of measure-preserving
transformations has been developed, and nothing interesting is likely to emerge from
these speculations.

To apply the technique to automorphism groups, it is necessary to consider the problem of choosing, for a given permutation group G, a random G-invariant graph. I'll
describe first the case when the group G is generated by one permutation which acts
as a single infinite cycle.

If the graph Γ admits the cyclic automorphism g, then the vertices can be labelled

$$\ldots, x_{-1}, x_0, x_1, x_2, \ldots,$$

so that g acts as the cyclic shift on the indices: $x_i g = x_{i+1}$ for all $i \in \mathbf{Z}$. Let S be the
set

$$\{\, i \in \mathbf{N} : x_i \sim x_0 \,\}$$

of positive integers. Then S determines Γ up to isomorphism, since $x_j \sim x_k$ if and
only if $|j - k| \in S$. It is also straightforward to show that S determines g up to
conjugacy in Aut(Γ); in other words, two cyclic automorphisms g and h of Γ are
conjugate in the group Aut(Γ) if and only if the corresponding sets $S(g)$ and $S(h)$
of positive integers are equal. Thus we have a bijection between pairs (Γ, g), where
Γ is a countable graph (up to isomorphism) and g is a cyclic automorphism of Γ
(up to conjugacy in Aut(Γ)), and subsets of $\mathbf{N} \setminus \{0\}$. Write $(\Gamma(S), g(S))$ for the pair
corresponding to $S \subseteq \mathbf{N} \setminus \{0\}$.

Now choose a set S of positive integers at random, by including positive integers
independently with probability $\frac{1}{2}$, as described in §1.4. As in (4.4), it can be shown
that, with probability 1, $\Gamma(S) \cong R$.

The argument is a little more complicated than before. We have to show that, once
finitely many points outside A have been chosen, we can find another point x outside
A so that decisions about edges and non-edges from x to A are independent of one
another and of all decisions previously made. Now the decisions about pairs $\{a, b\}$
and $\{c, d\}$ fail to be independent only if $|a - b| = |c - d|$, where we have identified

the vertex set with $\mathbf{Z}$. Now if x is chosen so large that $x - \max(A)$ is greater than the difference between any two points previously considered, then obviously x has the required independence.

A set with measure 1 certainly has cardinality $2^{\aleph_0}$. So we obtain the following result:

(4.5) Aut(R) *contains* $2^{\aleph_0}$ *pairwise non-conjugate cyclic permutations.*

One of the directions in which this result has been extended is the consideration of groups other than that generated by a single cyclic permutation. Given an arbitrary permutation group G on Ω, let $\mathcal{X}$ denote the set of orbits of G on 2-subsets of Ω. Then a G-invariant graph is obtained by choosing a subset of $\mathcal{X}$ (to be the edge set of the graph), and a random G-invariant graph by choosing a random subset of $\mathcal{X}$ (as usual, by including members of $\mathcal{X}$ independently with probability $\frac{1}{2}$).

I mentioned in the last section the determination by Truss (1985) of the cycle structures of automorphisms of R. This involves two parts: the easier part is to show that no other cycle structures occur, and the harder part is to show that all the claimed cycle structures really do occur. I proved the same result independently by showing that, for each "permissible" g, a random g-invariant graph is isomorphic to R with positive probability. (*Note*: We cannot say "with probability 1" here, Consider, for example, a permutation with one fixed point x and two infinite cycles C, D. We could choose to join x to neither, to both, or to just one of C and D; only in the last case — which has probability $\frac{1}{2}$ — do we have the possibility of getting a graph isomorphic to R, and in this case we do indeed obtain such a graph, apart from a null set.)

This suggests the following problem.

Problem. Is this statement true? Let G be any permutation group of countable degree. Then the following are equivalent:
  (i) some G-invariant graph is isomorphic to R;
  (ii) with positive probability, a random G-invariant graph is isomorphic to R.

Clearly, no other graph could have this property. As we've seen, it is true if G is a cyclic group. It is also trivially true if G is a group for which $f_2(G)$ is finite, since then the probability space is finite, and any isomorphism class which can occur will do so with positive probability. There is another case in which it holds, that in which G acts regularly (generalising the cyclic shift discussed above).

In a group G, a *square root set* is a set $\{x \in G : x^2 = a\}$ for some $a \in G$; it is *non-principal* if $a \neq 1$.

(4.6) *Let G be a countable group, and suppose that G is not the union of finitely many translates of non-principal square root sets. Then G is isomorphic to a regular subgroup of* Aut(R).

The proof follows that of (4.5) very closely. For the regular representation of G, pairs $\{a, b\}$ and $\{c, d\}$ lie in the same orbit if and only if $ab^{-1} = cd^{-1}$ or dc^{-1}. Now, if decisions about finitely many edges and non-edges have been made, we wish to choose an x so that the decisions about edges from x to A are independent of one another and all previous decisions. This involves avoiding solutions of $xa^{-1} = c$ for $a \in A$ and c in a finite set, and also solutions of $xa^{-1} = bx^{-1}$ for $a, b \in A$. The former excludes finitely many group elements. As to the latter, if $xa^{-1} = bx^{-1}$, then $(xa^{-1})^2 = ba^{-1}$, and so $x \in (\sqrt{ba^{-1}})a$, a translate of a non-principal square-root set. By hypothesis, there is an x avoiding all these solutions.

A group G is called a *B-group* if every primitive permutation group which contains the regular representation of G is 2-transitive. (In other words, if we adjoin to the right translations of G some permutations which destroy all the G-congruences, then all non-trivial binary relations are destroyed.) The B stands for Burnside, who first showed that finite cyclic groups of prime-power, non-prime order are B-groups. His methods and those of Schur, who extended this result to cyclic groups of arbitrary composite order, have been extremely influential in the development of finite permutation group theory. However, since the classification of the finite simple groups was completed in 1980, we know that most finite groups are B-groups, for a rather stupid reason: for a set of natural numbers n having density 1, the only primitive permutation groups of degree n are the symmetric and alternating groups.

By contrast, no countable B-group is yet known. The most powerful non-existence theorem is (4.6): since Aut(R) is primitive but not 2-transitive, any group satisfying the hypothesis of (4.6) is not a B-group. See Cameron and Johnson (1987) for discussion. I mention one further result in the same direction.

(4.7) *No countable abelian group is a B-group.*

For, in an abelian group G, the square roots of the identity form a subgroup A, and the non-empty square-root sets are cosets of A. If the hypothesis of (4.6) doesn't hold, then A has finite index in G, and so G has finite exponent (at most $2|G : A|$); thebn G is the direct product of two countable subgroups, and is isomorphic to a regular subgroup of $S \operatorname{Wr} \operatorname{Sym}(2)$ (with the product action).

It is possible to write down a necessary and sufficient condition in terms of translates of square root sets for a group G to act regularly on R. (This is just a matter of translating condition (♣) appropriately.) The condition is considerably more elaborate than the hypothesis of (4.6), but I don't know whether it is really more general.

Exercises

1. All the results in this section are unaffected if "probability $\frac{1}{2}$" is replaced with "probability p", where p is any fixed number strictly between 0 and 1.

2. A *digraph* is an irreflexive binary relation; an *oriented graph* is an irreflexive, antisymmetric binary relation. Write down conditions analogous to (♣) which characterise the countable homogeneous universal (a) digraph, (b) oriented graph, (c) tournament. Hence find sufficient (or, if you can, necessary and sufficient) conditions for the countable group G to act regularly on each of these objects. Are any new non-B-groups obtained?

3. Prove the assertion made in this section, that a random g-invariant graph, where g is a permutation of $\mathbf{N}$ with one fixed point and two infinite cycles, is isomorphic to R with probability $\frac{1}{2}$. Which other graphs occur with positive probability?

4.4. MEASURE, CONTINUED

The probabilistic technique described in the last section works well when structures of the type being considered can be described by a sequence of independent decisions. An obvious test case concerns $(k + 1)$-uniform hypergraphs, which generalise the situation already considered. As far as regular subgroups go, the position is even nicer than before. As in §2.6, let R_k denote the countable homogeneous universal $(k + 1)$-uniform hypergraph.

(4.8) *For $k \geq 2$, every countable group acts as a regular subgroup of* $\mathrm{Aut}(R_k)$.

The result corresponding to (♣) for R_k (the translation of (♠) of §2.8) asserts that, given a finite set A of vertices, and a set B of k-subsets of A, there is a vertex z such that, for a k-subset K of A, $K \cup \{z\}$ is a hyperedge if and only if $K \in B$.

Now it suffices to show that, given a finite set A, there exists a point z such that the G-orbits on $(k + 1)$-subsets of $A \cup \{z\}$ containing z are distinct from each other and from any given finite set of orbits, so that decisions about these orbits can be made independently. (Here G is an arbitrary countable group in its regular representation.)

Let O be an orbit of G on $(k+1)$-sets and $L \in O$. Then, for each $a \in A$, at most $k+1$ members of O contain a, namely, the sets $Lu^{-1}a$ for $u \in L$. It follows that sets in the finitely many prescribed orbits which meet A cover only a finite number of points.

On the other hand, if K_1 and K_2 are k-sets for which $K_1 \cup \{z\}$ and $K_2 \cup \{z\}$ lie in the same orbit, then for each $u \in K_1$, zw^{-1} has the form uv^{-1} for some $u, v \in K_2 \cup \{z\}$. For at least one choice of w, neither u nor v is equal to z, and so $z = uv^{-1}w$. Thus, given K_1 and K_2, there are only finitely many such z. (Here, we have used the fact that $k > 1$.)

So all but finitely many elements z have the required property.

Moreover, there is no need to stick to regular subgroups. Although no general analysis has been made, the following is an example of what can be done:

(4.9) *For any countable field F, the 3-transitive group* $\mathrm{PGL}(2, F)$ *is a subgroup of* $\mathrm{Aut}(R_3)$.

(An orbit of the group $\mathrm{PGL}(2, F)$ on 4-subsets of the projective line over F is specified by the cross-ratios of its members, a set of six (or, in special cases, fewer) elements of $F \setminus \{0, 1\}$ forming an orbit of the group

$$\langle x \mapsto 1 - x, \; x \mapsto 1/x \rangle.$$

Choose a random set of these cross-ratio orbits. The corresponding set of quadruple orbits almost surely form a hypergraph isomorphic to R_3.)

For most interesting structures, however, the choices are not independent, and the methods must necessarily be *ad hoc* (though the possible beginning of a general theory is sketched in §4.10). I'll consider just two special cases.

First, consider $\mathrm{Aut}(\mathbf{Q}, <)$. We cannot order a set by making decisions about the orders of pairs of points independently. But, in special cases, we can describe all the G-invariant orders. Consider the group $G = \mathbf{Z} \times \mathbf{Z}$, the free abelian group of rank 2. An order on G which is translation invariant is determined by the set of positive elements. This set is almost completely specified by a linear functional f on $\mathbf{R}^2$, by the rule that $(x, y) > (0, 0)$ if $f(x, y) > 0$. Also, f can be assumed to have norm 1, that is, to be represented by a point on the unit circle. If the ratio of the coordinates of f is irrational, then f completely determines the order, which is dense (without endpoints) and so is isomorphic to $\mathbf{Q}$. If, on the other hand, the ratio of the coordinates of f is rational (or infinite), then there are two possible orders; we have to specify which "end" of the line $f(x, y) = 0$ is to be positive. So the set of all G-invariant orders is bijective with the unit circle with its "rational" points doubled. If we use Lebesgue measure on the circle, the rational points form a null set, and so it holds that a random G-invariant order is almost surely isomorphic to $\mathbf{Q}$.

To introduce the next example, I will quote the classification of homogeneous countable graphs. The theorem below is due to Lachlan and Woodrow (1980); the examples under (ii) were found by Henson (1971).

(**4.10**) *A countable homogeneous graph is isomorphic to one of the following:*
    (i) the disjoint union of m complete graphs of size n, where at least one of m and n is infinite, or the complement of this graph;
    (ii) the countable homogeneous graph G_n whose age consists of all finite graphs containing no complete subgraph of size n ($n \geq 3$), or the complement of this graph;
    (iii) the "random graph" R.

Henson (1971) also showed that G_3 admits cyclic automorphisms, but G_n does not for $n \geq 4$. Can the positive result be proved by probabilistic methods?

Recall the correspondence of the last section between pairs (Γ, g), where Γ is a countable graph (up to isomorphism) and g a cyclic automorphism of Γ (up to conjugacy in $\mathrm{Aut}(\Gamma)$), and subsets of $\mathbf{N} \setminus \{0\}$. (Given $S \subseteq \mathbf{N} \setminus \{0\}$, the graph $\Gamma(S)$ has vertex set $\mathbf{Z}$, with $j \sim k$ if and only if $|j - k| \in S$; the automorphism g is the map $j \mapsto j+1$.) Now clearly $\Gamma(S)$ is triangle-free if and only if S is *sum-free*, that is, if $x, y \in S$ then $x + y \notin S$.

There is a simple recipe for a random sum-free set S of positive integers. Consider the positive integers in turn. For each n considered, if $n = j + k$ with $j, k \in S$, then by definition $n \notin S$; otherwise, choose with probability $\frac{1}{2}$ whether to put n into S, all such decisions being independent. Now the question is: Is it true that $\Gamma(S) \cong G_3$ with probability 1? The answer is negative. I suspect (but cannot prove) that the probability is actually zero. There are, however, infinitely many graphs which occur with positive probability. The first of these is the "universal almost homogeneous bipartite graph" (see below), whose probability is approximately 0.218.... .

A bipartite graph which is not null or complete bipartite cannot be homogeneous, since some non-edges consist of points in the same bipartite block while others contain points from different blocks. Call a bipartite graph *almost homogeneous* if, when equipped with a new binary relation whose interpretation is "lie in the same bipartite block", it is homogeneous. (In other words, we specify the bipartition, and require

that only those isomorphisms between finite subgraphs which respect this bipartition extend to automorphisms.) There is a unique countable almost homogeneous
bipartite graph B whose age consists of all finite bipartite graphs (with all possible
bipartitions); this is the graph referred to above. Now, if a set S consists of odd numbers only, then $\Gamma(S)$ is bipartite (with bipartition into the even and odd integers). It
can be shown that the three events

    (a) S consists of odd numbers,

    (b) $\Gamma(S)$ is bipartite,

    (c) $\Gamma(S) \cong B$,

differ pairwise by null sets, and their common probability is as claimed. See Cameron
(1987c), or for a broader picture, Cameron (1987a).

Exercises

1. We can specify a probability measure on the linear orderings of a countable set
by the rule that, given any n points, all of their $n!$ orderings are equally likely. Show
that, with this measure, the random order is almost surely isomorphic to $(\mathbf{Q}, <)$.

2. We can specify a probability measure on the set of triangle-free graphs on a given
countable vertex set as follows. Order the set of pairs of vertices (with order type
ω). Now consider the pairs of vertices in turn. If joining a pair would create a
triangle, then don't join; otherwise, decide at random (as usual, independently with
probability $\frac{1}{2}$). Find an ordering for which the random triangle-free graph is almost
surely isomorphic to G_3. (It is an open *problem* to characterise all the orderings for
which this holds. See Covington (1986).)

3. Show that a non-abelian free group admits a dense order which is invariant under
right translation. [*Hint*: the lower central factors are free abelian.] Hence show that
$\mathrm{Aut}(\mathbf{Q}, <)$ contains a non-abelian free subgroup.

Remark: A group admitting a translation-invariant order is obviously torsion-free;
but the converse is false. (For an example due to S. J. Pride, see Mura and Rhemtulla
(1977), p. 127.)

4. Show that the following procedure for choosing a random bipartite graph yields
the "almost homogeneous" graph B with probability 1:

Take the vertex set to be $\mathbf{N}$. For each vertex n in turn, choose whether to assign it
to the "left" or to the "right" bipartite block; then choose, for each vertex previously
assigned to the other block, whether to join it to n. (All choices independent with
probability $\frac{1}{2}$.)

Research problem. For which positive integers m, n does the group $\mathbf{Z}^m$ act regularly on the homogeneous universal n-tuple of linear orders?

4.5. CATEGORY

The method of Baire category is applicable to all situations where we already obtained positive results using measure in the last two sections. When it applies, it is usually easier than the probabilistic method. However, there is no obvious categorical analogue of such results as "among triangle-free graphs with cyclic automorphisms, a proportion 0.218... are isomorphic to the universal almost homogeneous bipartite graph B."

First, to reiterate the basics from §1.4. In a situation where an object is determined by an ω-sequence of choices, a set S of objects is open if, for any object $s \in S$, some initial segment of the choices determining s guarantees membership in S; it is dense if any initial segment of choices has a continuation in S; and it is residual if it contains a countable intersection of open dense sets. Residual sets are non-empty (indeed, dense and of cardinality $2^{\aleph_0}$), and a countable intersection of residual sets is residual.

As an illustration, here is the proof of the category version of the Erdős-Rényi theorem (4.4): a residual set of countable graphs consists of graphs isomorphic to R. Recall the property (♣) which characterises R. Now it suffices to show that the set S of graphs which satisfy (♣) for a fixed pair (A, B) of finite sets is open and dense, since (♣) is the intersection of all these sets and there are only countably many choices of A and B.

S is open. For if a graph belongs to S, then there is a vertex X whose neighbour set in A is precisely B; and then any graph for which the same choices about edges from x to A are made, also belongs to S.

S is dense. If finitely many decisions have been made, then there is a vertex x for which no decisions about edges from x to A have yet been made, so we are free to join x appropriately.

Logicians will recognise this technique as "finite forcing". I'll describe it, and the analogous measure-theoretic version, a little further in the appendix to this chapter (§4.10).

The measure-theoretic proof that R has many cyclic automorphisms also has its category-theoretic counterpart, as do its generalisations (to regular automorphism groups of R, and to the similar results for hypergraphs). In all these cases, both methods work for the same class of groups.

Remarkably, this method shows that Henson's universal homogeneous triangle-free graph G_3 has cyclic automorphisms: in the notation of the last section, a residual set of sum-free sets S satisfy $\Gamma(S) \cong G_3$. In this case, unlike the others we have considered, measure and category give quite different pictures of the "typical" sum-free set. (Sets containing only odd numbers have measure $0.218\ldots$ but are meagre.) Other aspects of this intriguing contrast are described in Cameron (1987a).

It is worth noting that there are many cases in which no isomorphism class is residual or of measure 1 (or even of positive measure). The most extreme example, that of N-free graphs, is taken from Covington (1989).

An *N-free* graph is one which doesn't contain the path of length 3 as an induced subgraph. Covington showed that there is a unique countable "almost homogeneous" N-free graph. To explain: If $\{a, b, c\}$ is a null subgraph of an N-free graph, then there cannot be vertices x, y whose neighbour sets in $\{a, b, c\}$ are different 2-subsets. So a null graph of order 3 contains (at least potentially) a distinguished vertex. We add to the language a ternary relation which distinguishes one vertex in each null subgraph of order 3. Dually, we do the same for each complete subgraph of order 3. (The class of N-free graphs is self-complementary.) For tidiness, we use the same ternary relation for both jobs, and also let it distinguish the one vertex in each non-complete non-null 3-set which is distinguished by the graph structure (that is, the mid-point of each path of length 2, and dually). It turns out that, over the extended language, there is a unique countable homogeneous object whose age consists of all finite structures arising from N-free graphs in the manner described.

It is worth noting that the automorphism group of the ternary relation alone is the 3-homogeneous, not 2-primitive group of (3.25), which is also the stabiliser of a point in the "boron trees" group. We'll return to this in §5.4.

This almost homogeneous graph, however, does not admit a cyclic automorphism. Indeed, in the space of pairs consisting of an N-free graph (up to isomorphism) and a cyclic automorphism (up to conjugacy), each isomorphism type occurs exactly once:

(4.11) *A countable N-free graph admits at most one conjugacy class of cyclic automorphisms.*

To see this, we identify points of the space with subsets S of $\mathbf{N}$, as usual. (The condition on S which is equivalent to $\Gamma(S)$ being N-free can be easily written down.)

Then we show that any such subset is determined, up to complementation, by a sequence $(n_1, n_2, \ldots)$ (finite or infinite) of integers greater than 1, as follows: assuming that $1 \in S$ (by complementation if necessary), then

S contains all positive integers not divisible by n_1;

S contains no positive multiples of n_1 not divisible by $n_1 n_2$;

S contains all positive multiples of $n_1 n_2$ not divisible by $n_1 n_2 n_3$;

and so on.

Now the graph determines the sequence recursively as follows: the complement of S has n_1 connected components, each associated with the sequence $(n_2, n_3, \ldots)$. (If the sequence is finite, the process terminates with an infinite complete graph.)

The preceding paragraph suggests a shift operator and the possible use of techniques of ergodic theory here. But it isn't clear to me what questions we should be trying to answer with these techniques!

Exercises

1. Let X be a group. For $a_1, a_2, \ldots, a_n \in X$, set

$$S(a_1, a_2, \ldots, a_n) = \{x \in X : a_1^{-1} x a_2^{-1} x \ldots a_n^{-1} x = 1\}.$$

Note that $S(a_1) = \{a_1\}$, and $S(a_1, a_2)$ is a translate of the square root set of $a_1^{-1} a_2$. Also, let

$$C(a, b) = \{x \in X : x^{-1} a x = b\},$$

a coset of the centraliser of a if a and b are conjugate, empty otherwise.

Prove the following. Let X be a countable group which cannot be expressed as the union of finitely many sets of the form $S(a)$, $S(a, b)$ $(a \neq b)$, $S(a, b, c)$, or $C(a, b)$. Then X acts regularly on the graph G_3. Moreover, if X is abelian, then $C(a, b)$ can be deleted from this list.

Which abelian groups satisfy this condition?

2. Prove that the countable elementary abelian 2-group acts regularly on Covington's N-free graph.

[The graph can be defined as follows. The vertex set is the set of all finite subsets of $\mathbf{Q}$. Choose a countable dense subset A of $\mathbf{Q}$ whose complement is also dense. Now, if $\{x_0, \ldots, x_{m-1}\}$ and $\{y_0, \ldots, y_{n-1}\}$ are distinct vertices with $x_0 < \ldots < x_{m-1}$ and $y_0 < \ldots < y_{n-1}$, suppose that $x_i = y_i$ for $i = 0, \ldots, k-1$ but not for $i = k$; without loss of generality, either $k = n < m$, or $k < \min(m, n)$ and $x_k < y_k$. Join the two vertices if and only if $x_k \in A$.

Now the vertex set is a group G under the operation of symmetric difference, indeed an elementary abelian 2-group. Show that, in its regular representation, G consists of graph automorphisms. It is easiest to show that taking the symmetric difference with a singleton set is an automorphism; the singletons generate G.

The most difficult part of the argument is to identify this graph with Covington's, that is, to show that, enriched with the ternary relation described in the text, it is homogeneous.]

4.6. MULTICOLOURED SETS

Let M be a homogeneous structure whose age has the strong amalgamation property. Then (3.9) gives a simple but powerful technique for finding subgroups of $\text{Aut}(M)$: simply impose on M another structure satisfying the same conditions, and there will be a subgroup preserving both structures.

This idea can be applied to produce unusual actions of the free group of countable rank. This depends on another easy group-theoretic technique due to Tits (1974), which can be formulated as follows.

(4.12) *Let G be a permutation group on a countable set Ω, and N a normal subgroup of G having the same orbits on Ω as G. Suppose that there is a subgroup H of G containing N such that H/N is free of countable rank. Then there is a subgroup F of G such that*
(i) $FN = H$, $F \cap N = 1$;
(ii) F is free of countable rank;
(iii) F has the same orbits on Ω as G.

To see this, let $f_0 N$, $f_1 N$ ... be free generators for H/N. Enumerate all pairs (α, β) of elements of Ω for which $\alpha g = \beta$ for some $g \in G$; say (α_0, β_0), (α_1, β_1), ... Since N has the same orbits as G, we can choose, for each i, an element $n_i \in N$ such that $\alpha_i f_i n_i = \beta_i$. Then the group F generated by the elements $f_i n_i$ for $i \in N$ obviously satisfies (iii). Its generators are coset representatives for the generators of H/N; so $FH = N$. Let $w(x_i)$ be a reduced word such that $w(f_i n_i) \in N$. Then $w(f_i N) = N$; so, by hypothesis, w is the trivial word. We conclude that $F \cap N = 1$ and that F is free on the given generators.

Before turning to the main results of this section, I give two applications of (4.12). The first is the highly transitive permutation representation of the free group F_ω of countable rank promised in §4.2; the second, the original application of the result by Tits.

(4.13) *The group F_ω has a faithful highly transitive permutation representation in which each non-identity element fixes only finitely many points.*

We apply (4.12) with $G = \mathrm{Sym}(\mathbf{N})$ and N the finitary symmetric group (the group of all permutations having finite support). Take

$$\Omega = \bigcup_{n \geq 1} \mathbf{N}^n;$$

the high transitivity of N guarantees that G and N have the same orbits on Ω. Take $H = XN$, where X is the regular representation of F_ω; then $X \cap N = 1$, so that $H/N \cong X$ by the third isomorphism theorem. The result follows. Every element of $H \setminus N$, and hence every non-identity element of F, is the product of a finitary permutation and a fixed-point-free permutation, and hence has only finitely many fixed points, as claimed.

We call a permutation *cofinitary* if it fixes only finitely many points, and say that a permutation group is *cofinitary* if all its non-identity elements are cofinitary.

The next application is more geometric in nature.

(4.14) *For every n, the group F_ω acts faithfully and flag-transitively on the n-dimensional projective space over some countable field.*

We take the field k to be a pure transcendental extension of a prime field with countable transcendence degree. Then take $G = \mathrm{P\Gamma L}(n+1, k)$, $N = \mathrm{PGL}(n+1, k)$; then $G/N = \mathrm{Aut}(k)$ contains the symmetric group on a transcendence basis, and hence the free group F_ω. Also, G and N are both flag-transitive. Now apply (4.12).

The main subject of this section is the following application, taken from Cameron (1989b).

(4.15) *Let M be a homogeneous structure whose age has the strong amalgamation property, and let k and l be integers with $k \leq l$. Then there is a subgroup F of $\mathrm{Aut}(M)$ with the properties:*
(i) F is free of countable rank;
(ii) F had the same orbits as $\mathrm{Aut}(M)$ on k-tuples, but not on $(k+1)$-tuples;

(iii) the maximum number of points fixed by a non-identity element of F is l.

I'll prove this first for $l = k$, and then indicate the simple modification required for the general proof. Take a countable set C of "colours", and let $\mathcal{C}$ be the class of finite structures in which the $(l+1)$-subsets are coloured with colours from this set, every $(l+1)$-set being given a unique colour. Then $\mathcal{C}$ has the strong AP; so there is a homogeneous structure M' whose age is $\text{Age}(M) \wedge \mathcal{C}$. We take $N = \text{Aut}(M')$, and G the group of permutations which preserve the structure of M' defined over the language of M and permute the colours among themselves.

Then $G/N \cong \text{Sym}(C)$. For the structure obtained from M' by re-labelling the colours according to an arbitrary permutation is homogeneous and has the same age as M', and so is isomorphic to M' (by (2.13)); this isomorphism is an element of G inducing the given permutation of C.

The structure obtained by restricting M' to the language of M is homogeneous and has the same age as M; so it is isomorphic to M, by (2.13) again. In particular, this implies that $G \leq \text{Aut}(M)$. We can identify the point set of M' with that of M.

Furthermore, since $\mathcal{C}$ gives no structure to a set of cardinality k or smaller, G and N have the same orbits on M^k as does $\text{Aut}(M)$.

Now let H be the subgroup of G inducing the regular representation of F_ω on C, and Ω the set M^k; and let F be the group obtained from (4.12). All is clear except (iii).

Since F permutes C regularly, an element of F fixing an $(l+1)$-tuple fixes its colour and so is the identity. The converse, viz. the existence of a non-identity element fixing l points, doesn't follow from the construction as given, but is achieved by a small modification. Use the generators $f_1 n_1$, $f_2 n_2$, ... in Tits' lemma to ensure the transitivity, reserving f_0 for another purpose. Now we can find points $x_1, \ldots, x_l, y, z$ such that the map θ fixing $x_1, \ldots, x_l$ and mapping y to z is an isomorphism of M-substructures, and such that the colour of $\{x_1, \ldots, x_l, z\}$ is the image of that of $\{x_1, \ldots, x_l, y\}$ under f_0; then we can choose $n_0 \in N$ such that $f_0 n_0$ extends θ. Thus $f_0 n_0$ is an element of F fixing l points.

To handle the case when $k < l$, we first enrich M by a universal homogeneous $(k+1)$-uniform hypergraph, to split the orbits of $\text{Aut}(M)$ on $(k+1)$-tuples; then proceed as before.

Is the hypothesis $k \leq l$ really necessary? It is easy to see that the result is false if $k > l - 1$; but the case $k = l - 1$ is open.

Problem. Given a countable homogeneous structure M whose age has the strong amalgamation property, and a positive integer k, is there a subgroup G of Aut(M) with the property that, given any two k-tuples of distinct elements lying in the same Aut(M)-orbit, there is a unique element of G carrying the first to the second?

The results in the last few sections of this chapter give an affirmative answer for many structures in the case when $k = 1$.

Consider the case where M is a set. A group with the property of the problem in this case is called *sharply k-transitive.* For $k = 1$, it is just a regular permutation group; and, as Cayley showed, there is no restriction at all on the isomorphism class of such a group. There are familiar examples for $k = 2$ (the affine group

$$\mathrm{AGL}(1, k) = \{x \mapsto ax + b : a, b \in k, a \neq 0\}$$

over any skew field k) and $k = 3$ (the projective group

$$\mathrm{PGL}(2, k) = \{x \mapsto \frac{ax + b}{cx + d} : a, b, c, d \in k, ad - bc \neq 0\}$$

over a commutative field k). But Tits (1952) showed:

(4.16) *For $k \geq 4$, there is no infinite sharply k-transitive permutation group.*

(4.15) has an unexpected application to the theory of scales of measurement in mathematical psychology. Stevens (1946) first put forward the point of view that a scale of measurement is a class of maps from some domain of objects to a number system (usually the real numbers), any two such maps being related by a restricted order-preserving permutation of the number system. The restriction on the order-preserving permutations is taken to be membership in a specified group. Thus it is this group, rather than the individual maps, which is important. For example, in the case of length, the number system is the positive reals, and multiplication by positive reals is allowed (corresponding to changing the unit of length). For temperature (before Kelvin, that is, without the concept of absolute zero), the number system is **R**, and the group is

$$\{x \mapsto ax + b : a, b \in \mathbf{R}, a > 0\}$$

(corresponding to choice of zero and unit). At the other extreme, there are many situations in which, although numerical measurements are made, only the ordering of the measurements is significant; the group here consists of all order-preserving permutations.

An important parameter of a scale is its *scale type* (m, n), where m is the degree of homogeneity (the greatest number such that the group is transitive on m-tuples in strictly increasing order), and n the degree of uniqueness (the least number such that the stabiliser of n distinct points is trivial). Otherwise put, m is the greatest value such that the numbers assigned to m objects can be assigned arbitrarily (subject only to their being in the correct order), and n the least value such that an assignment of numbers to n objects completely determines the measurement system. In the three examples in the preceding paragraph (which are historically those of the greatest importance), the scale type is $(1,1)$, $(2,2)$, and (∞, ∞) respectively.

Clearly any scale type (m, n) satisfies $m \leq n$. The best result on real scale types is due to Alper (1987):

(4.17) *If a real scale type (m, n) satisfies $m > 0$ and n finite, then $(m, n) = (1, 1)$, $(1, 2)$ or $(2, 2)$.*

Alper goes further and describes all possible groups having these scale types, up to conjugacy in the group of all order-preserving permutations of $\mathbf{R}$. Those of scale type $(1, 1)$ and $(2, 2)$ must be the ones given earlier. (The example for $(1, 1)$ is more conveniently described, by taking logarithms, as the group of translations of $\mathbf{R}$.) A group with scale type $(1, 2)$ lies between the two.

. On the other hand, for rational scale types, it follows immediately from (4.15) that the position is completely different:

(4.18) *For any m, n with $m < n$ (m, n finite), there is a rational scale with scale type (m, n).*

This may be a contribution to the argument about using real rather than rational numbers for measurement: if we used rationals, there would be too many possible scales with finite type.

The construction technique can also be applied to scale types with n infinite (where the group is cofinitary), and more generally to the corresponding problem in homogeneous structures whose age has the strong amalgamation property. See exercise 1 below, or Cameron (1989b).

Another application of the "multicoloured" technique of proof of (4.15), with some ingenious modifications, is due to Cohen (1986). He shows that, for every positive integer e, there is a perfect e-error-correcting code (of countable length) which has an e-transitive automorphism group.

Exercises

1. Construct, for any k, a faithful permutation representation of F_ω which is k-transitive (but not $(k+1)$-transitive), and is cofinitary but with no upper bound on the numbers of fixed points of non-identity elements.

[*Hint*: Use the embedding of F_ω in $\mathrm{Sym}(\mathbf{N})$ given by (4.13) in place of the regular representation.]

2. The question of scale type (k,k) with $k > 2$ in general is unresolved. Prove that, if there is a group of order-preserving permutations of any infinite linearly ordered set Λ with scale type (k,k) $(k > 2)$, then there is such a group with $\Lambda = \mathbf{Q}$.

4.7. ALMOST ALL AUTOMORPHISMS?

We have used the fact that, for certain groups G, the set of G-invariant structures of a certain type (graphs, triangle-free graphs, etc.) forms a complete metric space, in order to find G as a subgroup of the automorphism group of a specially nice structure of this type. But the automorphism group of an object is itself a complete metric space; can we turn the machine of Baire category onto it, perhaps to find other kinds of subgroups? Less is known here, but I hope to show that there are some promising lines of investigation.

In this section, I will consider only automorphism groups of homogeneous structures whose ages have the strong amalgamation property. For these groups, all the non-trivial orbits of the stabiliser of a tuple are infinite; so there is room to choose a point in a specified orbit avoiding any finite number of points. This will obviously be useful in constructing automorphisms satisfying specified constraints. However, it is not clear that this is the only, or even the best, class of closed permutation groups to which these speculations can be applied.

The ease with which we found cyclic automorphisms of the random graph R might suggest that the set of cyclic automorphisms is residual. But a moment's thought shows that this cannot be so, since this set is disjoint from the stabiliser of a point (an open set). Indeed, a dense set must contain permutations with arbitrarily many fixed points. Moreover, $\mathrm{Aut}(\mathbf{Q}, <)$ contains no cyclic automorphisms, and indeed no automorphisms with only finitely many cycles.

However, the set of automorphisms with only finitely many finite cycles is dense under our hypothesis:

(4.19) *Let $G = \text{Aut}(M)$, where M is homogeneous and $\text{Age}(M)$ has the strong amalgamation property. Then the set S of automorphisms with only finitely many finite cycles is dense in G.*

We have to show that any isomorphism between finite subsets of M can be extended to a member of S. Indeed, it can be extended in such a way that any finite cycle is already present in the initial data. We do this by extending the isomorphism (and its inverse) one point at a time, avoiding the finitely many choices which would close a cycle. According to the remarks above, this is always possible.

However, John Truss has shown that my intuition in searching for residual sets here was misguided. For several particular choices of M, he exhibits a residual set disjoint from the set S of (4.19). Let

$$T = \{g \in \text{Aut}(M) : g \text{ has infinitely many cycles of each finite length and no infinite cycles}\}.$$

In the case where M is a set, and $\text{Aut}(M)$ the symmetric group, T is a single conjugacy class, and Truss shows that it is residual. In the case where M is the random graph, he imposes a further condition, picking out a subset of T which is a single conjugacy class and again is residual.

Problem. For which structures M does $\text{Aut}(M)$ contain a *generic* conjugacy class, that is, one which is residual in $\text{Aut}(M)$? (Of course, there is at most one such class.)

I was led to consider the possibility that another possible property might be residual, from the following theorem of Samson Adeleke (1988):

(4.20) *Let H and K be countable cofinitary subgroups of $\text{Sym}(\Omega)$, where Ω is countable. Then there exists $g \in \text{Sym}(\Omega)$ such that H^g and K generate their free product, which is also countable and cofinitary.*

Here H^g denotes the conjugate $g^{-1}Hg$.

Problem. Let H and K be countable cofinitary subgroups of $\text{Aut}(M)$, where M is homogeneous and $\text{Age}(M)$ has the strong amalgamation property. Is the set

$$\{g \in \text{Aut}(M) : H^g \text{ and } K \text{ generate their free product, which is countable and cofinitary}\}$$

residual in $\text{Aut}(M)$?

I had no good evidence for thinking that this might be true. Its attraction was two-fold:

(i) If we can prove that things with certain properties exist (as Adeleke has done for us, in the case of the symmetric group), it often turns out that almost all things have those properties.

(ii) The problem suggests its proof strategy. Of the required conclusions, countability is of course trivial, while the other two come from the same source: we have to show that, if w is any non-trivial reduced word in the free product $H^g * K$, then w (interpreted as a permutation) is cofinitary (and so, necessarily non-trivial). Because the intersection of countably many residual sets is residual, and because there are only countably many such words w, it suffices to show that, for a fixed w, the set of elements $g \in \mathrm{Aut}(M)$ for which this holds is residual. It is only this "technical detail" which stands in the way of a proof. (Of course, this consideration applies with equal force to any weakening of the cofinitarity condition, such as assuming that every non-trivial word moves infinitely many points, or merely that every non-trivial word induces a non-trivial permutation.)

Unfortunately, this, too, has been refuted by John Truss. He showed that, if h and k are cofinitary permutations, then the set
$$\{g \in \mathrm{Sym}(\Omega) : g^{-1}hgk \text{ fixes at least } n \text{ points}\}$$
is open and dense in the symmetric group. The intersection of these sets, over all natural numbers, is thus residual and is disjoint from the set of the conjecture. However, it still seems possible that some weaker form of the conjecture (as outlined in the preceding paragraph) might be true.

(4.19) guarantees many cofinitary infinite cyclic subgroups. (Other, more restricted such subgroups come from the earlier sections of this chapter.) So, if the problem had had an affirmative answer, we would have had another technique for finding free subgroups of finite or countable rank.

Also, it may be possible to do more, using the fact that countable intersections of residual sets are residual. Suppose, for example, that we have a structure M and cofinitary automorphisms h and k such that, for a residual set of $g \in \mathrm{Aut}(M)$, the group $\langle h^g, k \rangle$ is dense. Then (modulo an affirmative answer to some form of the problem) it is also free of rank 2 for a residual set. There seems to be the possibility of re-establishing some of the results about dense free subgroups by this technique.

Indeed, this was the application which motivated Adeleke. If h_n denotes a permutation of the countable set Ω which has n fixed points and a single infinite cycle, then any subgroup generated by conjugates of h_0, h_1, ... is highly transitive. Hence:

(4.21) *Any countable cofinitary permutation group is a free factor of a countable cofinitary highly transitive group, where the other free factor is a free group of countable rank.*

An affirmative answer to the problem for Aut(M) would enable (4.21) to be generalised to Aut(M) from the symmetric group. (Use the cofinitary free subgroups given by (4.15) to produce a dense free subgroup; then one further application of the problem gives the result.

An interesting related question was posed by Tim Wall.

Problem. Is it possible to choose conjugates of h_0, ..., h_{n-1} (as defined above) so that the group they generate is
  (i) free of rank n,
  (ii) not $(n+1)$-transitive? (It is clearly n-transitive.)

4.8. SUBGROUPS OF SMALL INDEX

I begin this section with an apparently unrelated question. Let G be the automorphism group of an $\aleph_0$-categorical structure M, To what extent does G determine M?

The answer depends on how much we are prepared to assume about G. Three natural subproblems take as given the structure of G (1) as permutation group (2) as topological group (3) as abstract group.

(1) If we are given G as permutation group, then we know the canonical structure N for G. Now the relations, functions, etc. in M are unions of G-orbits, and so are definable in M; and since M is $\aleph_0$-categorical, the relations in N, which are G-orbits, are types in M. The situation is as satisfactory as it could be.

(2) A subgroup of G is open if and only if it contains the stabiliser of some tuple. For the stabilisers of tuples form a basis of neighbourhoods of the identity, and so if H is open then it contains one; and conversely, if H contains the stabiliser K of a tuple, then it is a union of cosets of K, each of which is open.

The structure of G as topological group determines the permutation representation on the union of the coset spaces of all the open subgroups. By the above paragraph, these are obtained by taking all orbits of G on tuples, and all quotients of these by congruences (G-invariant equivalence relations). It should be noted that this permutation representation still has countable degree; for G has finitely many orbits on n-tuples for each n, each orbit is countable, and each orbit has only finitely many congruences (since a congruence is a union of G-orbits on $2n$-tuples). Moreover, if N is any finite union of these orbits, then G has only finitely many orbits on n-tuples from N for all n; in other words, G acts oligomorphically on N.

The permutation representation described in the preceding paragraph is known to model theorists as M^{eq}. Thus, the structure of G as topological group determines M^{eq}, and M is some finite union of orbits in M^{eq}. (Not all finite unions of orbits are possible here; for example, it is necessary that G act faithfully on the chosen orbits, and that it be the full automorphism group of the corresponding canonical structure.)

In some cases, it is possible to select from M^{eq} a distinguished structure M, for example, by insisting that algebraic closure in M is trivial. I will not pursue this here.

(3) We cannot hope to obtain an answer in this case which is more precise than the one we obtained in case (2). That is, at best, M^{eq} could be determined by the abstract structure of G. So we could modify the question and ask: when does the structure of G as abstract group determine its topology?

We observed implicitly above that any open subgroup has finite or countable index. When, conversely, is it true that any subgroup having at most countable index is open (that is, contains the stabiliser of some tuple)? It has become customary to weaken "countable" to "less than $2^{\aleph_0}$" here. Accordingly, we say that G has the *small index property* if any subgroup of index less than $2^{\aleph_0}$ contains the stabiliser of a tuple; and G has the *strong small index property* if any subgroup of index less than $2^{\aleph_0}$ lies between the setwise and the pointwise stabiliser of a finite set. Also, we say that the $\aleph_0$-categorical structure M has the (strong) small index property if its automorphism group does.

(4.22) *A countable set with no additional structure, or a vector space of countable dimension over a finite field, has the strong small index property.*

The result for the symmetric group was announced by Semmes (1981), and a proof was published by Dixon, Neumann and Thomas (1986); the version for linear groups over finite fields is due to Evans (1986). I outline briefly a proof of the first result.

Let H be a subgroup of index less than $2^{\aleph_0}$ in $S = \mathrm{Sym}(\Omega)$. Then $H \cap G$ has index less than $2^{\aleph_0}$ in G, for any subgroup G of S. Take for G the stabiliser of countably many disjoint infinite sets whose union is Ω. Then G is the cartesian product of countably many copies of S, and it follows that H_Δ induces $\mathrm{Sym}(\Delta)$ on Δ for some infinite set Δ. (If $H \cap G$ induced a proper subgroup of $\mathrm{Sym}(\Delta)$ on every set Δ of the partition, then we could find $2^{\aleph_0}$ elements of G which lie in pairwise distinct cosets of H.)

Now any proper normal subgroup of $\mathrm{Sym}(\Delta)$ has index $2^{\aleph_0}$ in $\mathrm{Sym}(\Delta)$ (see the next section for this fact). It follows that the pointwise stabiliser of $\Omega \setminus \Delta$ in H induces the full symmetric group on Δ.

To conclude the argument, take a set A minimal with respect to the property that the pointwise stabiliser of A in $\mathrm{Sym}(\Omega)$ is contained in H. (We've shown that coinfinite sets with this property exist, and it is easily seen that if two sets have this property and their union is not the whole of Ω, then their intersection also has the property.) It is possible to show that A is finite, and that H fixes A setwise. This completes the proof.

It is easy to find examples having the small index property but not its strong version. The simplest is $S \mathrm{Wr} S$, in which the stabiliser of a congruence class has countable index but fixes no finite set. There are several further cases (due to Truss (1989b) and Evans (to appear)) in which the small index property holds:

(4.23) **Q** *(as ordered set), the countable atomless Boolean algebra, and vector spaces over finite fields carrying symplectic, orthogonal or unitary forms, all have the small index property.*

Note that, by (2.8), a *closed* subgroup of index less than $2^{\aleph_0}$ in $\mathrm{Aut}(M)$, for any $\aleph_0$-categorical M, contains the stabiliser of a tuple.

It was believed for a while that all $\aleph_0$-categorical structures might have the small index property. However, this is false. A counterexample was found by Hrushovski, and simplified by Hodkinson; it appears (in a slightly different guise) in a footnote (due to Cherlin) in Lascar (1982).

(4.24) *There is an $\aleph_0$-categorical structure M such that $\mathrm{Aut}(M)$ has a quotient which is the cartesian product of countably many copies of the cyclic group of order 2.*

Let R_k^* be the structure consisting of a complementary pair of hypergraphs each isomorphic to R_k, regarded as being interchangeable. (It can be described by a

$2(k+1)$-ary relation $R(\bar{x}, \bar{y})$ which holds when $\bar{x}$ and $\bar{y}$ are both $(k+1)$-tuples of distinct elements and both or neither are hyperedges in R_k.) The age of R_k^* has the strong amalgamation property. Take the conjunction (in the sense of (3.9)) of the ages of all these structures, and let M be the corresponding homogeneous structure. Then, in the automorphism group of M, we can swap the hypergraphs R_k with their complements for an arbitrary set of positive integers k, so the result holds.

Now the cartesian product of countably many copies of the cyclic group of order 2 has $2^{2^{\aleph_0}}$ subgroups of index 2. On the other hand, the stabiliser of an m-tuple contains elements interchanging R_k with its complement for an arbitrary subset of $\{k : k \geq m\}$, and so lies in only finitely many of these subgroups of index 2. So the vast majority of these subgroups are not open.

By modifying this construction, it is possible to obtain any profinite group with a countable basis of open subgroups as a (topological) quotient of the automorphism group of some $\aleph_0$-categorical structure (Evans and Hewitt (to appear)).

To conclude this section, I mention a result of Rubin (to appear). This gives a sufficient condition for an $\aleph_0$-categorical structure in which algebraic closures are trivial (that is, $\mathrm{acl}(F) = F$ for every finite set F) to be recoverable uniquely from its automorphism group (as abstract group). The condition holds in a wide class of homogeneous binary relational structures, of which the prototype is the random graph R.

A number of recent papers, by Macpherson and Neumann, Macpherson and Praeger, Macpherson and Thomas, and Thomas (all to appear), concern more general questions about subgroups of the symmetric group. One of the chief concerns is the question:

Which subgroups of $\mathrm{Sym}(\Omega)$ *are contained in maximal subgroups?*

For example, the main result of Macpherson and Praeger asserts that any subgroup which is not highly transitive is contained in a maximal subgroup. Other results concern maximal *closed* subgroups. Macpherson and Neumann prove that $\mathrm{Sym}(\mathbf{N})$ is not the union of a countable chain of proper subgroups (with the obvious extension to symmetric groups of larger degree). On the other hand, Thomas proves that the symmetric group S of degree λ has a subgroup G such that the subgroups lying between G and S form a well-ordered chain of length 2^λ (so that G lies in no maximal subgroup). Thomas' proof requires a group-theoretic hypothesis, which he proves to be consistent with but independent of ZFC (Zermelo-Fraenkel set theory with the Axiom of Choice).

4.9. NORMAL SUBGROUPS

There is not a great deal that can be said in general about normal subgroups of automorphism groups of $\aleph_0$-categorical structures. At one extreme, we have a result of Truss (1985):

(4.25) Aut(R) *is simple.*

Truss' result is better than indicated here; it applies to a more general class of structures, and it shows that, given non-identity elements g and h, g can be written as a product of five conjugates of h or h^{-1} (this is considerably stronger than simplicity).

At the other extreme, the group of (4.24) has $2^{2^{\aleph_0}}$ (normal) subgroups of index 2. (This of course is the maximum possible number of normal subgroups of a permutation group of countable degree.) This example is not homogeneous over a finite language; but this deficiency has been remedied by Droste, Holland and Macpherson (1989a),(1989b) who showed that the automorphism group of a countable homogeneous semilinear order (see §5.1) has $2^{2^{\aleph_0}}$ normal subgroups.

In between, we have two classical results, due to Schreier and Ulam (1933) and Higman (1954) respectively:

(4.26) *(i) The proper non-trivial normal subgroups of* Sym($\mathbf{N}$) *are the group of finitary permutations and the group of finitary even permutations (the alternating group).*
(ii) The proper non-trivial normal subgroups of Aut($\mathbf{Q}, <$) *are the group of order-automorphisms whose support is bounded below, the group of order-automorphisms whose support is bounded above, and their intersection (the group consisting of order-automorphisms with support bounded above and below).*

The normal subgroup lattices of these two examples are shown in Fig. 3.

Exercises

1. Show that the group C of permutations preserving the cyclic order on the roots of unity is simple.

2. (Truss). Find a non-trivial proper normal subgroup of the group of almost-automorphisms of the random graph R. [*Hint*: Consider the parity of the number of adjacencies changed by a permutation.]

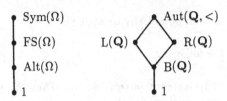

Fig. 3. Normal subgroup lattices

4.10. APPENDIX: THE TREE OF AN AGE

In this section, I'll give a somewhat more systematic account of the application of measure and category to structures with a given age.

Let M be any infinite relational structure. We regard the age of M as a tree, as follows. The nodes of the tree at level n are all the structures in $\text{Age}(M)$ whose point set consists of the first n natural numbers $0, \ldots, n-1$. The "parent" or immediate predecessor of a node at level n $(n > 0)$ is the induced substructure on the first $n-1$ natural numbers.

Now any infinite upward path from the root in this tree defines a relational structure N, over the same language as M, on the point set $\mathbf{N}$. (An instance of a relation holds in N if and only if it holds in the structure corresponding to any node on the path which contains all the points involved in the tuple.) Moreover, N is younger than M (that is, $\text{Age}(N) \subseteq \text{Age}(M)$), since any finite substructure of N is contained in the structure on the first n natural numbers for some n, which by definition belongs to $\text{Age}(M)$.

Conversely, if N is a structure on the point set $\mathbf{N}$ which is younger than M, then M is represented by a path in the tree. For the induced substructure of N on $0, \ldots, n-1$ belongs to $\text{Age}(M)$ and so is a node of the tree; and these nodes form a path.

Let $\mathcal{X}(M)$ denote the class of all structures on the point set $\mathbf{N}$ which are younger than N. By our remarks, $\mathcal{X}(M)$ is bijective with the set of paths in $\text{Age}(M)$ (regarded as a tree), and we will often identify these objects.

A path in the tree is defined by a sequence of decisions. The number of decisions to be made at each stage may depend on the path, and may be infinite, though it will always be finite if M is $\aleph_0$-categorical or if the language has only finitely many "essential" relations of each arity. So we have the set-up to which the methods of

Baire category or measure can potentially be applied.

The key observation is the following.

(4.27) *Let ϕ be an $(\forall\exists)$-sentence satisfied by M. Then the set of structures in $\mathcal{X}(M)$ which satisfy ϕ is residual in $\mathcal{X}(M)$.*

Let ϕ be the sentence $(\forall\bar{x})(\exists\bar{y})\psi(\bar{x},\bar{y})$. It suffices to show that, for each tuple $\bar{a}$ of natural numbers, the set of structures in $\mathcal{X}(M)$ in which $(\exists\bar{y})\psi(\bar{a},\bar{y})$ holds is open and dense, since there are only countably many choices for $\bar{a}$. This is a simple exercise in the interpretation of the conditions of openness and denseness.

As a consequence, the set of structures satisfying all the sentences in any countable set of $(\forall\exists)$-sentences valid in M is residual. In particular:

(4.28) *If M is homogeneous, then the set of structures isomorphic to M is residual in $\mathcal{X}(M)$.*

In §2.8, we gave first-order axioms (H1)–(H3) for an $\aleph_0$-categorical homogeneous structure. Without assuming $\aleph_0$-categoricity, condition (H2) (asserting that every finite set carries a structure in $\mathrm{Age}(M)$) isn't first-order; but it is not needed here, since it is a consequence of our restriction to structures younger than M. The other axioms are $(\forall\exists)$ sentences.

Measure is more problematical. For simplicity, suppose that each node has only finitely many children (immediate descendents) — that is, any n-element structure in $\mathrm{Age}(M)$ can be extended to an $(n+1)$-element structure in $\mathrm{Age}(M)$ in only finitely many ways. We could define a measure simply by making the children of each node all equally likely. But this would have the unwanted effect that the probability of a finite structure will depend on the order in which its vertices appear. See Fig. 4 for an example of this phenomenon for triangle-free graphs.

I proposed a program for getting round this problem in a "natural" way, but it has difficulties which have not yet been overcome.

We can define a measure on $\mathcal{X}(M)$ by giving, for each element A of $\mathrm{Age}(M)$, the probability that a given set of $|A|$ natural numbers carries a structure isomorphic to A (with a prescribed labelling). In other words, if $|A| = n$ (so that the point set of A is $0,\ldots,n-1$), we let $p(A)$ be the probability that, for any given distinct $m_0,\ldots,m_{n-1} \in \mathbf{N}$, the map $i \mapsto m_i$ ($i = 0,\ldots,n-1$) is an embedding of A into N.

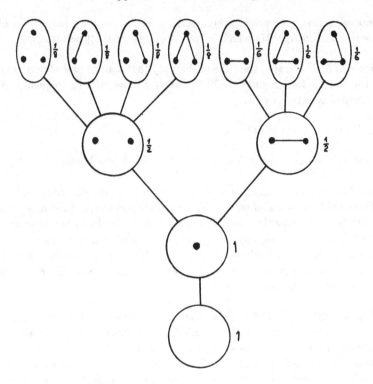

Fig. 4. The age of a triangle-free graph

It is easy to write down necessary and sufficient conditions on the values $p(A)$ for this rule to define a measure, namely

(M1) $p(A) \geq 0$ for all A;

(M2) $p(\emptyset) = 1$;

(M3) $p(A) = \sum p(A')$, where the sum is over the children A' of A;

(M4) $p(A)$ is independent of the labelling of A, i.e. p is isomorphism invariant.

It is interesting to note that conditions (M3) and (M4) comprise a system of linear equations for the values of p on $(n+1)$-element structures in terms of its values on n-element structures, and that the linear algebra proof of (3.2) is exactly what is needed to guarantee that these equations always have a solution. However, it is not obvious that the inequalities (M1) can be guaranteed.

The defect of the general approach, however, is that there are too many solutions, and they have no obvious connection with the structure under consideration. To rectify this, recall the heuristic justification of probability as limiting frequency. In accordance with this, I proposed taking $p(A)$ to be the limiting proportion of large labelled structures in Age(M) which induce A on the first $|A|$ points.

For example, if M is a graph, then the probability that a given pair of vertices should be joined in the random member of $\mathcal{X}(M)$ would be the limiting frequency of edges in large labelled subgraphs of M.

Problem. Do these limits exist?

Even in the simple case of edge frequency, the answer is not known.

If this problem has an affirmative answer, then these limits do indeed satisfy conditions (M1)–(M4) above, and so define a probability measure. In simple cases, the measure exists and is exactly what we expect. For example, in the random graph, every labelled n-vertex subgraph has probability $2^{-\binom{n}{2}}$, which is exactly the same as in our earlier probability measure for this graph. Similar assertions hold for the random $(k+1)$-uniform hypergraph, or the random total order of Exercise 1 of §4.4. But there are some surprises.

If a random triangle-free graph is chosen according to this method, then it is almost surely bipartite (and, indeed, is almost surely the "almost homogeneous" bipartite graph of §4.4). This follows from a result of Erdős, Kleitman and Rothschild (1976) on counting finite triangle-free graphs. They showed that, asymptotically, almost all are bipartite. This quite different sense of the phrase "almost all" is exactly what is needed here. For, if A is a finite triangle-free but non-bipartite graph, then the proportion of large triangle-free graphs which contain A tends to 0, and so $p(A) = 0$. So the probability that A is induced on a fixed subset is zero. Since there are only countably many subsets of size $|A|$, the probability that A is embeddable in the random graph N is zero. Again, since there are only countably many finite non-bipartite graphs, and a countable non-bipartite graph must contain one of them, the probability that N is non-bipartite is zero.

However, there are some positive aspects. For example:

(4.29) *Let M be a structure having the property that a random member of $\mathcal{X}(M)$ is almost surely isomorphic to M. Then the structure obtained from M by deleting finitely many points is isomorphic to M.*

Some time ago, I proposed calling a countable structure M *ubiquitous in category* (resp. *in measure*) if almost every structure younger than M is isomorphic to M. Thus, for example, every homogeneous structure is ubiquitious in category, while the random graph and $(\mathbf{Q}, <)$ are ubiquitous in measure. At the same time, I proposed calling M *absolutely ubiquitous* if every structure with the same age as M is isomorphic to M. (This notion implies ubiquity in category, at least — see Exercise 1.) Hodkinson and Macpherson (1988) showed that any relational structure which is absolutely ubiquitous is "trivial", in the sense that the domain can be partitioned into finitely many parts such that any permutation fixing these parts setwise is an automorphism. (The disjoint union of a finite graph, an infinite star graph, and an infinite complete bipartite graph would be a reasonably typical example.)

Macpherson went on to investigate other (not necessarily relational) structures, showing (for example) that an absolutely ubiquitous group must be abelian-by-finite. (For non-relational structures, the age is the class of all finitely-generated substructures; if we assume that the structure is $\aleph_0$-categorical, then the n-generator substructures are all finite, and have cardinality bounded by a function of n.)

Exercises

1. Show that the set of structures in $\mathcal{X}(M)$ having the same age as M is residual in $\mathcal{X}(M)$.

2. Show that, if M is the two-way infinite path, then the set of structures in $\mathcal{X}(M)$ which are isomorphic to M is residual. (This is true even though M is not characterised within $\mathcal{X}(M)$ by $(\forall\exists)$ sentences.)

3. Among partial Steiner triple systems, Steiner triple systems are residual. (A *partial Steiner triple system* consists of a set of points with a collection of distinguished 3-subsets called "triples", any two points lying in at most one triple; it is a *Steiner triple system* if any two points lie in exactly one triple.)

4. Show that, if M is a universal bipartite graph, then the measure on $\mathcal{X}(M)$ defined in the text is the same as the following. First, consider finite graphs with a fixed bipartition. To choose one at random, assign vertices randomly to the two bipartite blocks (independently with probability $\frac{1}{2}$, as usual); then choose randomly whether to join points in different bipartite blocks. Finally, coalesce isomorphic graphs with different bipartitions (such graphs must be disconnected).

By the result of Erdős, Kleitman and Rothschild, the same measure applies to triangle-free graphs (all non-bipartite graphs having probability zero). Determine the numbers which are assigned to the graphs in Fig. 4 by this measure.

5. Let M be a countable homogeneous relational structure. Suppose that there is a

function $g : \mathbf{N} \to \mathbf{N}$ such that any node at level n in the tree of Age(M) has exactly $g(n)$ children. Suppose further that $\sum 1/g(n)$ diverges. Prove that a random member of $\mathcal{X}(M)$ is almost surely isomorphic to M.

Verify the hypotheses in the case where M is $(\mathbf{Q}, <)$.

Remark: The condition that $\sum 1/g(n)$ diverges is not necessary for the conclusion. It fails for the random graph, where $g(n) = 2^{n+1}$, and for the universal pair of linear orders (Example 3 of §3.3), where $g(n) = (n+1)^2$; in both of these cases, the conclusion holds. *Subsidiary Exercise*: Prove these assertions; *Problem*: Strengthen Exercise 5 to cover them.

6. Show that an absolutely ubiquitous relational structure over a finite relational language is $\aleph_0$-categorical.

5

Miscellaneous topics

5. Miscellaneous topics

5.1. JORDAN GROUPS

Let G be a permutation group on Ω. A *Jordan set* for G is a subset of Ω with the property that the pointwise stabiliser of its complement acts transitively on it. (Sets consisting of just one point satisfy this condition trivially but are usually excluded for technical reasons.) If G is n-transitive, then any set containing all but $n-1$ points of Ω is a Jordan set; such Jordan sets are called *improper*. (This needs some care in the case when n is infinite.) Then G is called a *Jordan group* if it has a proper Jordan set (other than the empty set).

With the exception of some recent examples constructed by Hrushovski (to appear), the known infinite Jordan groups are of three types:

(J1) Geometric examples: These are the projective group $\mathrm{PGL}(n,k)$, the affine group $\mathrm{AGL}(n,k)$, and their close relatives. The pointwise stabiliser of any subspace of a projective or affine space acts transitively on its complement. So the complements of subspaces are the Jordan sets, and the geometry can be recovered from them. In this class, it is customary now to include also the automorphism groups of algebraically closed fields (which preserve the geometry of algebraically closed subfields). In each of these cases, the subspaces of the geometry are precisely the algebraically closed sets (in the sense of §2.7). This fact is crucial, both in their study, and in applications.

(J2) Automorphism groups of relational structures based on various kinds of orders or trees. For example, $\mathrm{Aut}(\mathbf{Q},<)$ is a Jordan group; any open interval (finite or semi-infinite) is a Jordan set. I'll discuss these examples further later.

(J3) Groups of homeomorphisms of certain topological spaces such as manifolds. Such groups are usually highly transitive, and so are outside the scope of our discussion; but one example is discussed in §5.5. (Another countable example is **Q**, as topological space, which has already cropped up several times; see Neumann (1985a).)

There are three main areas in which classification theorems for infinite Jordan groups have been obtained. The first concerns finite Jordan sets. But a much weaker hypothesis suffices:

(5.1) *An infinite primitive permutation group containing a non-identity element of finite support contains the alternating group (and so is highly transitive).*

A proof of this theorem was outlined in Exercise 3 of §2.7. Neumann (1975), (1976) has extended the result to a powerful theory of finitary permutation groups.

The second is in a sense "dual" to the first, concerning groups with a cofinite Jordan set. The result in question is more-or-less implicit in Cherlin-Harrington-Lachlan (1985) (for reasons I'll try to explain in §5.3), and explicit in Neumann (1985b):

(5.2) A primitive permutation group having a cofinite Jordan set is either highly transitive, or a subgroup of a projective or affine group over a finite field.

Note that G is highly transitive if and only if every cofinite set is an (improper) Jordan set.

The proof depends on the way in which the geometry associated with such a Jordan group lends itself to induction.

A *pregeometry* (often called a matroid or combinatorial pregeometry) consists of a set X with a closure operator cl on the power set of X satisfying
  (i) $Y \subseteq \text{cl}(Y)$;
  (ii) $Y \subseteq Z$ implies $\text{cl}(Y) \subseteq \text{cl}(Z)$;
  (iii) (the *Exchange Axiom*) if $y \in \text{cl}(Y \cup \{z\})$ but $y \notin \text{cl}(Y)$, then $z \in \text{cl}(Y \cup \{y\})$;
  (iv) the closure of a set is equal to the union of the closures of its finite subsets.
It is a *geometry* (or geometric matroid) if the empty set and all singletons are closed.

From any pregeometry, we can obtain a geometry in a canonical way by deleting the closure of the empty set, and then identifying points whose closures are equal. (Having the same closure is an equivalence relation.)

The subspaces of a pregeometry are the sets which are closed (equal to their closures). A pregeometry is "homogeneous" if the pointwise stabiliser of a subspace is transitive on its complement. Thus, the automorphism group of a homogeneous pregeometry is

a Jordan group — the Jordan sets are the complements of the subspaces. The converse
is true if the geometry is finite. (This was explicitly pointed out by Marshall Hall
(1960), though it is implicit in Jordan's work.) Furthermore, the converse holds if the
geometry is *locally finite*, that is, if every finite set is contained in the complement
of a cofinite Jordan set, see Neumann (1985b).

A *disintegrated* geometry is one in which every set is closed. The automorphism
group of a disintegrated geometry is the symmetric group.

(**5.3**) An infinite, locally finite, homogeneous geometry is either disintegrated, or
a projective or affine geometry over a finite field.

As noted, the proof depends on inductive properties of such geometries. In a homogeneous pregeometry, the restriction of the closure operator to any closed set defines
a homogeneous subgeometry; quotient geometries can also be defined, and are also
homogeneous. Now a homogeneous locally finite geometry is in a sense built from
homogeneous finite geometries, and these can all be classified; all those of sufficiently
large rank are either disintegrated or projective or affine spaces.

A word about the proof is in order. The automorphism group of any homogeneous
geometry is 2-transitive. The classification of the finite simple groups leads to a
complete list of the finite 2-transitive groups, and it is not too hard to decide which
of these can act on geometries: there are only the types mentioned and a few small
exceptions (of which the most interesting are the large Mathieu groups.) The result
is very much in the spirit of finite group theory. As is well known, the finite simple
groups fall into infinite families apart from twenty-six "sporadic" ones.

Subsequently, "elementary" proofs, not using the classification of finite simple groups,
have been given by Zil'ber (1988) and Evans (1986a); their methods are geometric
and combinatorial. They are able to show that a finite homogeneous geometry of
sufficiently large rank (at least seven in Zil'ber's best result) having more than two
points on a line is projective or affine. (The *rank* of a geometry is one less than the
number of subspaces in the longest chain; *points* and *lines* are subspaces of rank 1
and 2 respectively.) The arguments used have enriched both model theory and finite
geometry; but it must be said that the strongest result (the determination of all finite
primitive Jordan groups) requires the classification of finite simple groups.

The third type of characterisation is contained in a major piece of work by Adeleke
and Neumann (three papers, to appear soon). This work achieves
   (a) the axiomatisation of the various classes of relational structures which arise
(generalised betweenness and separation relations, etc.);

(b) the determination of all the structures satisfying the axioms;

(c) a demonstration that any Jordan group satisfying some extra hypotheses is an automorphism group of such a structure.

A *semilinear order* is a partial order satisfying the conditions

(a) every two elements have an upper bound;

(b) for all x, $\{y : y \geq x\}$ is linearly ordered.

(It is an object which looks something like a tree, with the root at the top; but the branching need not occur on discrete levels, so it could resemble a thorn bush rather than a tree.)

A semilinear order is *proper* if it is not a linear order (that is, some pair of elements is incomparable).

All countable homogeneous semilinear orders have been determined (Droste (1985)). Such a structure is determined by two parameters, the (finite or countable) *ramification number*, the number of "branches" below a point, and whether or not *least* upper bounds exist in axiom (a).

For example, let Ω denote the set of all non-empty finite sequences

$$\alpha = (q_0, q_1, \ldots q_{n-1})$$

of rational numbers. Write $\alpha < \beta$ if either

(i) $\alpha = (\sigma, q)$ and $\beta = (\sigma, r, \tau)$, where σ and τ are arbitrary sequences (possibly empty) and $q < r$, or

(ii) α is a proper initial subsequence of β.

It is readily checked that $(\Omega, \leq)$ is a semilinear order with ramification number 2 and having least upper bounds.

On any semilinear order, it is possible to define a ternary *generalised betweenness* relation, motivated by our picture of the semilinear order as a treelike object. If α and β are comparable, then γ lies between them if it does so in the order. If not, then γ lies between α and β if and only if it is above one of them but not above their supremum (in the Dedekind cut sense — that is, for every δ, if $\alpha \leq \delta$ and $\beta \leq \delta$ then $\gamma \leq \delta$. See Fig. 5(a).) Imagine picking up the tree by the end of any infinite branch, and putting this end at the top; the result is another semilinear order giving rise to the same generalised betweenness relation. Because of this, the automorphism group of the generalised betweenness relation may be larger than that of the semilinear order, and in particular, may be 2-transitive.

The last two relations, chain relations and direction relations, present some additional complications. The set of all *chains* (maximal linearly ordered sets) in a semilinear order carries a natural ternary relation called a *chain relation*, which holds on the triple (C_1, C_2, C_3) of chains if and only if $C_2 \cap C_3 \supseteq C_1 \cap C_2 = C_1 \cap C_3$. (See Fig. 5(b).) However, interesting countable semilinear orders contain uncountably many chains, so we don't immediately get the structures we want. We can pick out a suitable countable subset of chains from an explicit description of the semilinear order, or apply the Löwenheim-Skolem theorem, or simply take the countable homogeneous structure with the same age as the uncountable structure we have constructed.

Similarly, a *direction relation* is a quaternary relation defined on the set of "ends of chains" (including the root) in a semilinear order, as illustrated in Fig. 5(c), or a countable analogue.

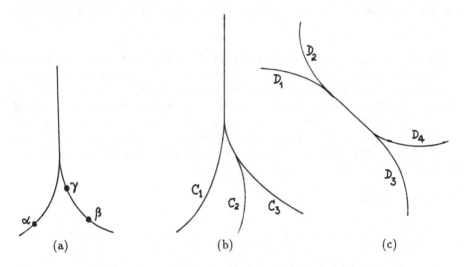

Fig. 5. Betweenness, chain and direction relations

The main result in the second paper of Adeleke and Neumann is the following.

(5.4) *Suppose that G is primitive but not highly transitive on Ω, and that G has a proper Jordan set Δ such that the pointwise stabiliser of $\Omega \setminus \Delta$ acts primitively on Δ. Then G preserves a linear order, betweenness, circular order, separation relation, or a proper semilinear order, generalised betweenness, chain, or direction relation on Ω.*

In each case, additional information is given about the action of G.

Exercise

Show that the "boron trees" group of §2.6 preserves a direction relation, while the stabiliser of a point in this group preserves a chain relation. (To do this thoroughly, it is necessary to use the axiom systems for these relations given by Adeleke and Neumann.)

5.2. GOING FORTH

In this section, I will investigate the back-and-forth argument a bit further, and consider the question of Adrian Mathias concerning the sufficiency of going forth only.

The showpiece application of back-and-forth today is the proof of Cantor's theorem characterising the rational numbers as countable dense linearly ordered set without endpoints. In fact, Cantor did not use back-and-forth; his argument is outlined below. Back-and-forth entered the mathematical mainstream with the book of Hausdorff (1914), but had been used earlier, in an exposition of Cantor's work by Huntingdon (1904). It seems probable that Huntingdon invented it. (These remarks are based on research by Jack Plotkin; I am grateful to him, and to Susan Schuur and Wilfrid Hodges for pointing me in the right direction here.)

Cantor's argument goes as follows.

Let $A = \{a_0, a_1, \ldots\}$ and $B = \{b_0, b_1, \ldots\}$ be two countable dense linearly ordered sets without endpoints. We construct a map $\phi : A \to B$ as follows. At stage 0, we set $a_0\phi = b_0$. Suppose that, after $n - 1$ stages, ϕ has been defined on $a_0, \ldots, a_{n-1}$, and is order-preserving on its domain. Then the points $a_0, \ldots, a_{n-1}$ divide A into $n + 1$ "minimal" intervals (including two semi-infinite intervals at the ends). Let (a_i, a_j) be the minimal interval containing a_n. (The notation suggests that it is a finite interval, but the argument goes in exactly the same way if it is a semi-infinite interval.) Then, since B is dense and without endpoints, the interval $(a_i\phi, a_j\phi)$ is non-empty; let b_m be the first point (in the given enumeration of B) which lies in this interval, and set $a_n\phi = b_m$. Clearly ϕ is still order-preserving on its extended domain.

After countably many stages, we have defined a map $\phi : A \to B$ which is one-to-one and order-preserving. I claim that ϕ is onto. Suppose not, and let b_p be the first point (in the enumeration) which is not in the range of ϕ. Then choose q such that

$$\{b_0, \ldots, b_{p-1}\} \subseteq \{a_0, \ldots, a_{q-1}\}\phi.$$

Then b_p lies in some minimal interval defined by the points $a_0\phi, \ldots, a_{q-1}\phi$, say $(a_i\phi, a_j\phi)$ $(i, j < q)$. Let a_r be the first point (in the enumeration of A) in (a_i, a_j). (This exists because A is dense and without endpoints.) Then, by definition, $a_r\phi = b_p$, contrary to assumption.

So ϕ is an isomorphism from A to B.

Note that the first paragraph of the proof, which only assumes that B is dense and without endpoints, shows that A is embeddable in B. This gives the result that $\mathbf{Q}$ is universal for countable linearly ordered sets.

Now we return to the general case. Recall the situation in which back-and-forth works, from §§2.5, 2.8. We have some notion of "type", which is an equivalence relation on n-tuples for each $n \in \mathbf{N}$ (the tuples are possibly chosen from different structures in some class), which satisfies

($\Diamond$) If $\mathrm{tp}(\bar{a}) = \mathrm{tp}(\bar{b})$ and x is any point in the structure containing $\bar{a}$, then there is a poiny y in the structure containing $\bar{b}$ such that $\mathrm{tp}(\bar{a}, x) = \mathrm{tp}(\bar{b}, y)$.

As we observed in Chapter 2, this guarantees that all the structures in the class are isomorphic (so we lose nothing by assuming that we have just one structure A), and also that tuples in A have the same type if and only if there is a type-preserving permutation of A which carries one into the other (so that the undefined "types" are just orbits of a closed group of permutations of A).

So we might as well dispense with structure and logic, and start with a closed permutation group G (on a set Ω), defining types to be its orbits on tuples. (These are just the isomorphism types, or the first-order types, in the canonical relational structure for G, if some structure is thought necessary.) Then the types satisfy ($\Diamond$).

Following Cantor's argument above, then, let $(a_0, a_1, \ldots)$ and $(b_0, b_1, \ldots)$ be two enumerations of Ω, and G a closed permutation group on Ω. *Types* are defined, as above, as orbits of G on $\bigcup_{n \geq 1} \Omega^n$. The *map defined by going forth* (with respect to the two enumerations) is the map ϕ specified recursively in the following manner:

If ϕ has been defined on $a_0, \ldots, a_{n-1}$, then $a_n\phi$ is defined to be the first point b_m (in the second enumeration) for which

$$\mathrm{tp}(a_0\phi, \ldots, a_{n-1}\phi, b_m) = \mathrm{tp}(a_0, \ldots, a_{n-1}, a_n).$$

(Such a point is guaranteed to exist by ($\Diamond$) and the fact (easily proved by induction) that ϕ is type-preserving.)

The map defined by forth is a one-to-one type-preserving map from Ω to Ω. We say that *forth suffices for G* if ϕ is always onto (irrespective of the chosen enumerations). We also say that *forth suffices for a structure M* if forth suffices for its automorphism group.

By way of contrast, here, in detail, is an example for which forth does not suffice. The structure in question is a countable dense ordered set without endpoints and with a distinguished dense subset whose complement is also dense. Such a structure is $\aleph_0$-categorical and homogeneous, and is naturally associated with the stabiliser of a point in the countable homogeneous local order (see §3.3). (It is interesting to note that, when I needed to show its uniqueness — a fact due, incidentally, to Skolem (1920) — and was unaware of back-and-forth, I used Cantor's argument but had to modify the map ϕ infinitely often to obtain the required isomorphism; see Cameron (1981).)

The proof strategy is to show that, given any second enumeration, it is possible to choose a first enumeration in such a way that the map defined by going forth is not onto. So let $(b_0, b_1, \ldots)$ be any enumeration. We will arrange that $b_0 \notin \mathrm{Im}(\phi)$. Note that, as soon as we have specified $a_0, \ldots, a_{n-1}$, then ϕ is defined on them; we assume that this has been done in such a way that $b_0 \notin \mathrm{Im}(\phi)$. We call the points in the distinguished set *red*, the other points *blue*.

Let b_k be the first point (in the second enumeration) which has not yet been put into the first enumeration. There is a unique minimal interval $(a_i\phi, a_j\phi)$ which contains b_0. If $b_k \notin (a_i, a_j)$, or if b_k lies in this interval but has the opposite colour to that of b_0, then we can set $a_n = b_k$ and guarantee that $a_n\phi \neq b_0$. So suppose that $b_k \in (a_i, a_j)$ and that b_k and b_0 have the same colour.

In this case, let b_m be the first point in $(a_i\phi, b_0)$ having the opposite colour to b_0, and c any point (without loss of generality, the first) in the interval (b_k, a_j) having the same colour as b_m. ("First" refers to the b enumeration.) Then set $a_n = c$, $a_{n+1} = b_k$; we have $a_n\phi = b_m$, and b_k is safely isolated from b_0. (See Fig. 6.)

The procedure ensures that every b_i is eventually put into the first (a) enumeration, and that b_0 never gets put into the image of ϕ.

I do not have necessary and sufficient conditions for forth to suffice. There is still a considerable gap between (5.7) and (5.9) below, though it seems that we can usually

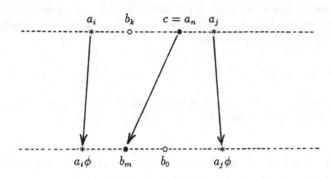

Fıg. 6. Isolating a point

decide by *ad hoc* methods in any particular case.

Borrowing a term from finite permutation group theory (but giving it a wider meaning), I define a *suborbit* of G to be a pair $(\bar{a}, A)$, where $\bar{a}$ is a tuple and A an orbit of the stabiliser of $\bar{a}$. (In the case where G is the automorphism group of an $\aleph_0$-categorical structure M, A is a minimal $\bar{a}$-definable set in M.) The suborbit $(\bar{a}, A)$ is *trivial* if A is a singleton which is a member of $\bar{a}$. Also, I say that $(\bar{b}, B)$ *dominates* $(\bar{a}, A)$ if

(i) each element of $\bar{a}$ occurs in $\bar{b}$;

(ii) $B \subseteq A$.

The suborbit $(\bar{a}, A)$ is called *splittable* if there is a tuple $\bar{b}$ of points outside A such that $G_{\bar{a}\bar{b}}$ is intransitive on A; it is *unsplittable* otherwise. Now, very loosely, if there are many unsplittable suborbits, then forth suffices; if there are many splittable orbits, then forth does not suffice. Before stating the results more precisely, I mention a couple of facts about these concepts.

(5.5) *Suppose that* $\mathrm{acl}(X) = X$ *for every finite set* X, *and let* $(\bar{a}, A)$ *be a splittable suborbit with the property that* $G_{\bar{a}}$ *acts primitively on* A. *Then every non-trivial suborbit dominating* $(\bar{a}, A)$ *is splittable.*

(5.6) *Suppose that* $(\bar{a}, A)$ *is a suborbit of the closed permutation group* G. *Then the following are equivalent:*

(a) every suborbit dominating $(\bar{a}, A)$ *is unsplittable;*

(b) for every suborbit $(\bar{b}, B)$ *dominating* $(\bar{a}, A)$, B *is a singleton or a Jordan set.*

For (b) clearly implies (a); and, if (a) holds, then for any permutation h induced on B by $G_{\bar{b}}$, we can simultaneuosly approximate h on B and the identity on $\Omega \setminus B$ arbitrarily closely by an element of G.

I come now to the main results.

(5.7) *Suppose that $(\bar{a}, A)$ is a suborbit with the property that every non-trivial suborbit dominating it is splittable. Then forth does not suffice.*

The proof of this follows very closely the argument given for the rationals with a dense, co-dense subset. (In that case, for any non-trivial suborbit $(\bar{a}, A)$, A consists of all the rationals of one colour in a minimal interval defined by $\bar{a}$; it can be split by any point of the other colour in the same interval.)

As a corollary of (5.5) and (5.7) we have:

(5.8) *Suppose that $\mathrm{acl}(X) = X$ for every finite set X, and that G has a primitive splittable suborbit. Then forth does not suffice.*

This result handles the random graph, Henson's homogeneous graphs G_n, the hypergraphs R_k, and so on.

In the other direction, we have:

(5.9) *Suppose that $(\bar{a}, A)$ is a non-trivial suborbit with the property that, for every tuple $\bar{b}$ extending $\bar{a}$, there is a tuple $\bar{c}$ extending $\bar{b}$ such that $(\bar{c}, C)$ is unsplittable for all suborbits $(\bar{c}, C)$ dominating $(\bar{a}, A)$. Then forth suffices.*

As a corollary:

(5.10) *Under the equivalent conditions of (5.6), forth suffices.*

In particular, forth suffices if every orbit of the stabiliser of a tuple is a Jordan set (or a singleton). Peter Neumann (personal communication) has studied such groups, as a spin-off from his work with Samson Adeleke on Jordan groups described in §5.1. Let us say that a permutation group G has *property $\mathcal{P}$* if every orbit of the pointwise stabiliser of any finite tuple is a Jordan set (possibly a singleton or improper). Neumann shows that a closed group with property $\mathcal{P}$ is the direct product of finitely many transitive closed groups with property $\mathcal{P}$ (and conversely). Furthermore, the property is preserved by taking stabilisers of tuples or forming wreath products. Thus our interest focuses on primitive groups with property $\mathcal{P}$. (There is a gap here, which

Neumann discusses in detail.) Neumann shows that the primitive groups with property $\mathcal{P}$ preserve linear or circular orders, betweenness or separation relations, chains or direction relations, affine spaces over finite fields, or projective spaces over GF(2), or are highly transitive. (All the types listed here give examples.)

Another corollary of (5.9) is:

(5.11) *If G is closed and countable, then forth suffices.*

For there is a tuple whose stabiliser is trivial; and a singleton orbit is clearly unsplittable.

Exercises

1. Prove (5.5), (5.6), and (5.11) (the last to be done without (5.9)).

2. Decide whether forth suffices for your favourite structure or permutation group.

5.3. $\aleph_0$-CATEGORICAL, ω-STABLE STRUCTURES

This is an area where model theory and permutation group theory have come very close together, to their mutual benefit. Unfortunately for the non-logician, a substantial background of stability theory is required here. I shall not attempt to give this — I don't even give the definition of ω-stability — referring instead to Pillay (1983). The brief outline I present assumes $\aleph_0$-categoricity; this simplifies many definitions, and readers should be warned that much of what I say is false without this assumption. The presentation is strongly influenced by the lectures given by Simon Thomas at the Durham symposium.

A complete theory with no finite models is called *λ-categorical* (where λ denotes an infinite cardinal number) if any two models of the theory of cardinality λ are isomorphic. Morley (1965) proved that, if Σ is λ-categorical for one uncountable cardinal λ, then it is λ-categorical for every uncountable λ. We say that a theory is *uncountably categorical* if this holds, and *totally categorical* if it is also $\aleph_0$-categorical.

The theory of dense linear orders without endpoints is countably but not uncountably categorical. (The first assertion is Cantor's theorem. For the second, construct 2^λ models of cardinality λ for any uncountable cardinal λ as follows: start with λ, regarded as an ordinal, that is, a well-ordered set of cardinality λ; replace each element of λ with a copy of $\mathbf{Q}$; and add initial elements to the copies of $\mathbf{Q}$ at an arbitrary subset of the limit ordinals.)

On the other hand, the theory of free abelian groups is uncountably but not countably categorical. (A free abelian group is determined by a single invariant, its rank, which can be any cardinal number. If the rank is uncountable, then it is equal to the cardinality of the group; but if the rank is finite or countable then the cardinality is countable.)

As these examples may suggest, while $\aleph_0$-categoricity implies much symmetry of the countable model, uncountable categoricity implies something like a "structure theory" for all models, including the countable ones. It turns out that there is a wider class of theories to which the tools developed to study uncountable categoricity can be applied, namely the ω-stable theories. References for what follows include Zil'ber (1979), Cherlin-Harrington-Lachlan (1985), Hrushovski (to appear).

Let M be $\aleph_0$-categorical. A *definable set* in M is a union of orbits of the stabiliser of a tuple. The *Morley rank* of a definable set is given by the following specification. (David Evans points out to me that this is really Cantor-Bendixson rank, but in the $\aleph_0$-categorical case they are the same.)

  (a) A non-empty finite set has rank 0.
  (b) If X contains infinitely many pairwise disjoint definable sets of rank at least n, then X has rank at least $n + 1$.
  (c) If X has rank at least n but doesn't have rank at least $n + 1$, then X has rank n.
Normally, a definition like this would be continued into the transfinite; but one of the remarkable discoveries of Cherlin Harrington and Lachlan (1985) is that, if M is $\aleph_0$-categorical and has a rank, then it has finite rank. Moreover, ω-stability is equivalent to the condition that M has finite rank (as always, under the assumption of $\aleph_0$-categoricity).

The *degree* of a definable set of rank n is the maximum number of pairwise disjoint definable sets of rank n that it contains. (By definition, we cannot find infinitely many; then the fact that the number is bounded is proved by a compactness argument.)

Thus, any $\aleph_0$-categorical, ω-stable structure contains a set of rank 1 and degree 1. This is an infinite definable set having the property that any definable subset of it is either finite or cofinite. The first part of the argument is to determine such sets, which are called *strongly minimal*. Now a strongly minimal set X has the following property, where G is the group induced on X by the stabiliser of the tuple defining it:

(†) The stabiliser in G of any finite tuple in X has just one infinite orbit, which contains all but finitely many points; and this orbit is a Jordan set.

The fact that the stabiliser H of a tuple has just one infinite orbit is a consequence of the preceding remarks about strongly minimal sets. $\aleph_0$-categoricity shows that H has only finitely many orbits altogether, so the union of its finite orbits is finite. Now the stabiliser of all the points in these finite orbits is a subgroup K of H of finite index, so that the points in the infinite H-orbit lie in infinite K-orbits also; and K has only one infinite orbit, so it is transitive on the points it doesn't fix.

It follows easily from (†) that algebraic closure defines a locally finite, homogeneous pregeometry on X (as defined in §5.1). We proceed as there to pass from a pregeometry to a geometry, by removing a finite set (the closure of the empty set, which is the union of the finite orbits of G in X) and factoring out an equivalence relation with finite classes (identifying points with the same closure). The resulting set $\bar{X}$, on which G acts primitively, is called *strictly minimal*, and carries a locally finite homogeneous geometry. From (5.3), we obtain:

(5.12) *A strictly minimal set in an $\aleph_0$-categorical, ω-stable structure is either disintegrated (a set without any additional structure), or a projective or affine space over a finite field. The group induced on it is the symmetric group (in the disintegrated case) or contains the projective or affine group (in the other case).*

The ambiguity left by the last sentence concerns field automorphisms, in the case where the finite field is not a prime field.

The second question is: To what extent do the strictly minimal sets control the entire structure? This was settled in the case where $\text{Aut}(M)$ acts primitively by Cherlin, Harrington and Lachlan (1985). The result is a kind of coordinatisation theorem. The simplest case to explain is that in which all the strictly minimal sets are disintegrated.

(5.13) *Let M be a primitive $\aleph_0$-categorical ω-stable structure, all of whose strictly minimal sets are disintegrated. Then there are positive integers r and k, infinite sets $X_1, \ldots, X_r$, and a bijection between M and the set of r-tuples $(K_1, \ldots, K_r)$, where K_i is a k-element subset of X_i for $i = 1, \ldots, r$, such that the direct product of the symmetric groups on $X_1, \ldots, X_r$ (each symmetric group acting on the set of k-sets in the natural way, and the product acting on M via the bijection) is contained in $\text{Aut}(M)$. Moreover, $\text{Aut}(M)$ is contained in $\text{Sym}(X_1) \operatorname{Wr} \text{Sym}(r)$.*

In the general case, to formulate the coordinatisation theorem, we must replace the set of k-subsets of the infinite set by a *Grassmanian*, an orbit of $\text{Aut}(X)$ on finite algebraically closed subsets of X, where X is an infinite set or an infinite-dimensional projective or affine space over a finite field. There are two additional complications. The first is caused by the possibility of field automorphisms; the second, by the fact that the projective group is a homomorphic image of the affine group. See Cherlin-Harrington-Lachlan (1985) for a precise statement.

Without assuming primitivity, the glueing together of the strictly minimal sets is a far more complicated business: see Hrushovski (to appear).

A generalisation of the results of Cherlin, Harrington and Lachlan was obtained by Kantor, Liebeck and Macpherson (1989). Their hypotheses are abstracted from the way in which an infinite-dimensional projective space is built from its finite-dimensional subspaces.

Let N be a finite substructure of the $\aleph_0$-categorical structure M. (Use the canonical language for M, so that, in particular, it is homogeneous.) We say that N is a *finite homogeneous substructure* of M provided that two tuples in N lie in the same $\text{Aut}(N)$-orbit if and only if they lie in the same $\text{Aut}(M)$-orbit. (Note: since M is homogeneous, any automorphism of N is induced by an automorphism of M.)

Now M is *smoothly approximated* if it is the union of a chain of finite homogeneous substructures. There is an equivalent logical formulation: M is smoothly approximated if every sentence true in M is true in some finite homogeneous substructure of M. This shows, incidentally, that smoothly approximated infinite structures cannot be finitely axiomatised. Cherlin, Harrington and Lachlan (1985) showed that any $\aleph_0$-categorical, ω-stable structure is smoothly approximated.

Kantor, Liebeck and Macpherson determine all the primitive, smoothly approximated structures, and give information about the transitive ones. The primitive structures turn out again to be products of Grassmanians, but with a larger class of structures in place of the sets and the projective and affine spaces. The main difference is that the vector spaces are allowed to carry sesquilinear forms of classical type (symplectic, orthogonal or unitary); there are a couple of non-classical examples too.

The proof depends on the observation that, for any finite homogeneous substructure N of M, the number $F_n(\text{Aut}(N))$ of orbits of $\text{Aut}(N)$ on n-tuples of distinct elements is uniformly bounded by $F_n(\text{Aut}(M))$. All sequences of permutation groups for which such bounds exist can be determined (apart, of course, from a finite number of terms). In fact, using the classification of finite simple groups, Kantor, Liebeck and Macpherson show the following result:

(5.14) *There is a function $g : \mathbf{N} \to \mathbf{N}$ such that all finite permutation groups G whose degree n is greater than $g(F_5(G))$ are explicitly known.*

This of course generalises the determination of the finite 5-transitive permutation groups (the groups with $F_5(G) = 1$). Kantor, Liebeck and Macpherson remark that 5 can be reduced to 4 in this result.

Before leaving this section, I should mention a celebrated conjecture of Lachlan (1974), to the effect that any stable, $\aleph_0$-categorical structure is ω-stable. (Stability, in an $\aleph_0$-categorical structure, means that there is no definable relation $R(\bar{x}, \bar{y})$ and infinite set $\{\bar{a}_i : i \in I\}$ of tuples with the property that the set of pairs (i, j) of elements of I for which $R(\bar{a}_i, \bar{a}_j)$ holds is a linear order of I.) Lachlan showed that the obstructions to his conjecture are $\aleph_0$-categorical *pseudoplanes*, in the sense that an equivalent form of the conjecture is the non-existence of $\aleph_0$-categorical pseudoplanes. (A *pseudoplane* is an incidence structure of points and blocks, having the properties
  (i) any point is incident with infinitely many blocks, and any block with infinitely many points;
  (ii) any two points are incident with finitely many blocks, and any two blocks with finitely many points.)

It is a source of embarrassment to finite geometers that we do not even know whether or not there exists an $\aleph_0$-categorical projective plane. (A *projective plane* is a structure satisfying condition (ii) above with "finitely many" replaced by "exactly one" in both places.) Such a projective plane would have infinite chains of finite subplanes, many of which would have large automorphism groups.

In 1988, Hrushovski disproved Lachlan's conjecture by exhibiting an $\aleph_0$-categorical pseudoplane. In fact, he gave an uncountable family, parametrised by a real number α chosen from a residual subset of the interval $[\frac{1}{2}, \frac{2}{3}]$. These examples are based on graphs $G(\alpha)$ whose ages consist of all finite graphs Γ satisfying

$$|E(\Gamma)| \le f_\alpha(|V(\Gamma)|),$$

where f_α is a suitably chosen piecewise linear function. ($V(\Gamma)$ and $E(\Gamma)$ are the vertex and edge sets of Γ.) The graph $G(\alpha)$ is not homogeneous, but its age satisfies enough instances of the amalgamation property for the orbits of $\mathrm{Aut}(G(\alpha))$ to be described.

In §5.5, I will give an example of a structure (due to Wiegand (1978)) which satisfies exactly half of the definition of an $\aleph_0$-categorical pseudoplane: any block is incident with infinitely many points, any two blocks are incident with finitely many points, and the automorphism group is highly transitive on points.

5.4. AN EXAMPLE

It was not known to me for some time whether the information contained in the sequence (f_n), or even the cycle index, of an oligomorphic permutation group is enough to determine whether the group is primitive. The example in this section answers these two questions negatively.

The ingredients in this example have appeared separately at various points. I will link them up with a different presentation.

We start with the "boron trees" group of §2.6. Let G be the stabiliser of a point in the corresponding homogeneous structure. Then G is itself the automorphism group of a homogeneous structure M. The age of M consists of the following objects. The point set consists of all the H atoms in a boron tree except for a distinguished one r (which we regard as the "root"); it carries a ternary relation R such that $R(a, b, c)$ holds if and only if r, a, b, c stand in the original quaternary reltation. In other words, $R(a, b, c)$ holds if and only if the paths from the root to b and c agree on a longer initial segment than either does with the path from the root to a. This relation is a choice function distinguishing one point of each 3-element subset. (See Fig. 7.)

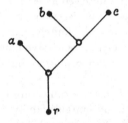

Fig. 7. A ternary relation

G is the 3-homogeneous but not 2-primitive group of (3.25). (For any point a, the relation defined by $b \sim c \iff R(a, b, c)$ is a G_a-congruence.)

The relation R is a chain relation, in the sense of §5.1, albeit a rather special one. I use the term "chain relation" below specifically to mean either this relation, or a member of its age.

Now enrich the age with extra structure, as follows. Colour the B atoms of the tree with two colours, black and white. (To continue the chemical metaphor, these are two isotopes of boron.) Now, given any two non-root H atoms x and y, there is a unique B atom $f(x,y)$ on the path joining them which is nearest to the root. Now form a graph whose vertices are the non-root H atoms, by joining x to y if and only if $f(x,y)$ is black.

The class of structures (graph and chain relation) defined in this way has the amalgamation property (and, trivially, satisfies the rest of Fraïssé's conditions). Let C be the countable homogeneous structure with this age, K the automorphism group of C, and H the group which preserves the chain relation and fixes or complements the graph (that is, which preserves or interchanges the colours of the B atoms).

The graph constructed here is Covington's almost homogeneous N-free graph (Covington (1989)), referred to in §4.5. The chain relation is determined by the graph, by the rules explained there. (If $\{a, b, c\}$ carries a subgraph which is neither complete or null, then one of the three vertices is distinguished by the graph structure; put $R(a, b, c)$ if a is distinguished. If $\{a, b, c\}$ carries a null graph, then there is a unique 2-subset of $\{a, b, c\}$ which is the neighbour set in $\{a, b, c\}$ of some vertex, by universality; put $R(a, b, c)$ if this subset is $\{b, c\}$. Do the dual of this if $\{a, b, c\}$ is complete.)

It is also informative to compare the formal definition of the relation R here with that of a chain relation in §5.1.

(5.14) *The groups K and $H \operatorname{Wr} G$ have the same cycle indices, and hence the same sequences (f_n) and (F_n); but the first is primitive and the second imprimitive.*

To see this, we have to find a bijection between the ages of the structures in question such that corresponding structures have automorphism groups which are isomorphic (as permutation groups). We must look a little more closely at the age of the $H \operatorname{Wr} G$-structure.

The graphs in the age of C are, as claimed, N-free, that is, contain no path of length 3 as induced subgraph. (This can be seen by inspection of a small finite number of cases.) Moreover, any N-free graph can occur here.

The class of finite N-free graphs is the smallest class of graphs which contains the 1-vertex graph and is closed under isomorphism, complementation, and disjoint union. It has the further property that the complement of a (finite) connected N-free graph is disconnected (and conversely, but the converse is trivially true for all graphs).

Suppose that the graph in one of our structures is disconnected. Then the following holds. If the points a, b, c lie in different components, then the truth of $R(a, b, c)$ is unaffected if any of these points is replaced by another point in the same component. So a chain relation is induced on the set of components. Moreover, if two of the three points lie in the same component, then the chain relation distinguishes the third point. An analogous assertion holds for connected graphs, using instead the partition of the structure into connected components of the complementary graph.

The age of the structure associated with $H \operatorname{Wr} G$ consists of structures of the following kind. The point set is partitioned, and the set of parts of the partition carries a chain relation. Each part of the partition carries a complementary pair of N-free graphs and a compatible chain relation. (This is immediate from the definition of the wreath product.) By our remarks above, this data specifies a chain relation on the whole set; instances of the relation where not all arguments belong to the same part of the partition are given by the rules in the preceding paragraph.

Now the bijection works as follows. Each age has just one 1-element structure, so we let these correspond. For any larger cardinality, the N-free graphs come in complementary pairs, with the same automorphism group (and each graph not isomorphic to its complement). We map the connected graph to the $H \operatorname{Wr} G$-structure having a single part, that part carrying the given complementary pair of N-free graphs and the given chain relation. We map the disconnected graph to the $H \operatorname{Wr} G$-structure for which the partition is into the connected components of the graph, and the chain relation on the set of parts is induced in the manner described above, the complementary pair of N-free graphs with compatible chain relation on each part being just the induced substructure from the given graph.

The primitivity of K is proved in the usual way, from the facts that K is transitive on edges and non-edges of the graph C and that the age of C contains all possible 3-vertex subgraphs, so that the diameter of C and its complement are both 2.

Note, in passing, that $f_n(H \operatorname{Wr} G) = 2 f_n(H)$ for $n > 1$.

Many elaborations of this construction are given in Cameron (1987b), which is a crib for the following exercise.

Exercise

Find transitive permutation groups G satisfying each of the following conditions:
  (i) $f_n(G \operatorname{Wr} S) = 2 f_n(G)$ for $n > 1$;
  (ii) $f_n(S \operatorname{Wr} G) = 2 f_n(G)$ for $n > 1$.
[Here S is the infinite symmetric group.]

5.5. ANOTHER EXAMPLE

This example is a slight generalisation of one due to Wiegand (1978). Take $k \geq 1$.
A *Wiegand k-structure* is an incidence structure of points and blocks which satisfies
the following three conditions:
  (W1) Any k blocks are incident with infinitely many points.
  (W2) Any $k + 1$ blocks are incident with only finitely many points.
  (W3) Let S be any finite set of blocks, T any finite set of points. Then there is a
block w such that
   (i) $T \subseteq w$;
   (ii) for any distinct $s_1, \ldots, s_k \in S$,

$$s_1 \cap \ldots \cap s_k \cap w \subseteq T.$$

(In (W3), and in the sequel, we identify a block with the set of points incident with
it. Clearly different blocks are incident with distinct sets of points.)

 (5.15) *For any $k \geq 1$, there is, up to isomorphism, a unique countable Wiegand
k-structure. Its automorphism group is highly transitive on points, and k-transitive
but not $(k + 1)$-transitive on blocks.*

This is due to Wiegand for $k = 1$; the general proof below is almost exactly the
same. Wiegand's motivation to look at these structures comes from the following
remarkable fact:

 (5.16) *The incidence structure whose points and blocks are the points and (irre-
ducible) algebraic curves in the affine plane over the algebraic closure of any finite
field is a Wiegand 1-structure.*

I will not give Wiegand's proof of this, which uses some algebraic geometry. (Note
that (W2) is Bezout's theorem.)

(5.16) shows that the structure obtained does not depend on the finite field chosen.
We also see that the automorphism group of the countable Wiegand 1-structure is a
Jordan group, since it contains the affine group. (The complement of any line, and
hence of any block, is a Jordan set.) Incidentally, the complement of any block x
in this structure, together with the blocks disjoint from x, is a Wiegand 1-structure,
and hence isomorphic to the original structure.

Finally, note that there is a topology (the Zariski topology) on the point set, for
which the proper closed sets are the sets consisting of a finite union of blocks and

finitely many more points. The topology determines the incidence structure, since blocks are the minimal infinite closed sets. Thus the automorphism group is the same as the homeomorphism group.

Proof of (5.15). The construction is very easy: start with the empty set, and alternately adjoin a block incident with each set of points, and a point incident with each k-set of blocks.

Given a finite set of points and blocks, its *closure* is obtained by adjoining all points incident with some $k + 1$ of the blocks; it is still finite. Note that any set of points, and any k-set of blocks, are closed. So the result follows by back-and-forth from the following assertion:

  ($\ddagger$) Let X and Y be Wiegand k-structures, A and B finite closed substructures of X and Y, $\phi : A \to B$ an isomorphism, and x an element of $X \setminus A$. Then there exists $y \in Y \setminus B$ such that ϕ extends to an isomorphism from $\mathrm{cl}(A \cup \{x\})$ to $\mathrm{cl}(B \cup \{y\})$ mapping x to y.

To show this, suppose first that x is a point. Then $A \cup \{x\}$ is closed. Since A is closed, x is incident with at most k blocks of A, say $b_1, \ldots, b_t$ $(t \le k)$. Choose $k - t$ blocks outside B incident with all points of B, say $c_1, \ldots, c_{k-t}$. Then choose y incident with $b_1\phi, \ldots, b_t\phi, c_1, \ldots, c_{k-t}$ but with no further block of B.

Now suppose that x is a block. Use the argument of the preceding paragraph to add, one at a time, the finitely many points of the closure of $A \cup \{x\}$ to A; so we may suppose without loss that $A \cup \{x\}$ is closed. Choose

$$S = \{a\phi : a \in A, \ a \in x\},$$

T the set of blocks of B together with k further blocks incident with every point of B, and let y be the block w guaranteed by (W3).

No doubt, there are further interesting structures waiting to be discovered here, either satisfying axioms like (W1)–(W3), or as configurations arising in algebraic geometry; and many more interesting properties of the Wiegand structures and their groups to be found.

5.6. OLIGOMORPHIC PROJECTIVE GROUPS

There is a fruitful and long-standing analogy between sets and projective spaces (especially over finite fields). Recall that a projective space is the lattice of subspaces of a vector space. For example, the modular geometric lattices are just direct sums of Boolean lattices, projective spaces, and non-Desarguesian projective planes. Moreover, many counting formulae applicable to projective spaces over the finite field $GF(q)$ become equal to the corresponding formulae for finite sets when q is put equal to 1. (The simplest example is the *Gaussian coefficient* $\begin{bmatrix} n \\ k \end{bmatrix}_q$, the number of k-dimensional subspaces of an n-dimensional vector space over $GF(q)$; letting $q \to 1$, we obtain $\binom{n}{k}$, the number of k-subsets of an n-set.)

It is natural, then, to call a group of collineations of an infinite dimensional projective space over a finite field *oligomorphic* if it has only finitely many orbits on the set of n-dimensional subspaces for each natural number n.

It is clear, incidentally, that an oligomorphic projective group acts as an oligomorphic permutation group on the points of the projective space (or, indeed, on the subspaces of any fixed dimension). However, it is more natural in this context to replace n-sets and n-tuples of distinct elements by n-dimensional subspaces and linearly independent n-tuples in the counting formulae.

For example, the formula given by Cameron and Taylor (1985) relates the numbers of orbits on linearly independent tuples with the numbers of orbits on all tuples, and is the analogue of (2.3).

Most of the results on this subject are contained in two papers by Simon Thomas (1986), (1988), in which he develops analogues of the content of the series by Cameron (1978), (1981), (1983a,b). I will give a survey of some of the highlights.

The analogue of the monotonicity result (3.2) is valid:

(5.17) *A projective group has at least as many orbits on $(n+1)$-spaces as on n-spaces.*

Both the proofs of (3.2) can be adapted. The linear algebra proof works exactly as for sets; the relevant finite "technical lemma" was proved by Kantor (1972). Ramsey's theorem for projective spaces was proved by Graham, Leeb and Rothschild (1972); the analogue of (1.10) is in Thomas (1988).

Curiously, it is not known whether (5.17) is valid for infinite fields.

The analogue of (3.10) is in Thomas (1988). To describe it, I must first explain a remarkable example constructed in that paper.

(5.18) *For any finite field* $GF(q)$, *there is an ordering of the point set of the projective space of countable dimension over* $GF(q)$ *with the following properties:*
  (i) the ordering is dense and without endpoints;
  (ii) for any finite n, *the unique order-preserving bijection between any two* n-*subspaces is induced by an order-preserving collineation of the whole space.*
Moreover, the structure is unique up to order-preserving collineation.

We conclude that the space is order-isomorphic to $\mathbf{Q}$ and is homogeneous, and that the automorphism group is transitive on n-spaces for all n. Thomas' proof was an application of Fraïssé's theorem; the amalgamation property is not entirely trivial to verify. (Note that a more general version of Fraïssé's theorem is being applied here: not every subset of a projective space carries a subspace. The important property is *local finiteness*: every finite subset is contained in a finite-dimensional subspace.) The uniqueness is immediate from the uniqueness part of Fraïssé's theorem.

Another, more simple-minded construction was given by Cameron and Hall (to appear). Their example carries a partial rather than a total order.

Let F be a field, and Λ a totally ordered set. Let $F[\Lambda]$ denote the F-vector space with basis Λ. (Since Λ may itself be a field, it is convenient to denote by $\tilde{\lambda}$ the basis vector corresponding to λ.) The *highest basis vector* of a vector $v \in F[\Lambda]$ is the largest element of Λ occurring with non-zero coefficient in v. Ordering vectors according to their highest basis vectors defines a partial order, for which "equal or incomparable" is an equivalence relation. A *special transvection* is a linear transformation which adds to a given basis vector $\tilde{\lambda}$ an arbitrary linear combination of smaller basis vectors, fixing all basis vectors except $\tilde{\lambda}$. The special transvections preserve the partial order, and generate a group H whose orbits are the incomparability classes.

The H-orbit of any finite-dimensional subspace contains a unique member which is spanned by basis vectors from Λ. Thus, if G is any group of order-preserving permutations of Λ (regarded as a linear group on $F[\Lambda]$ in the obvious way), then the number of orbits of the product $H.G$ on n-spaces is equal to the number of orbits of G on n-subsets of Λ. In particular, if we take $\Lambda = \mathbf{Q}$ and $G = \mathrm{Aut}(\mathbf{Q}, <)$, then $H.G$ is transitive on n-spaces for every n.

By taking a smaller group G, we can construct groups with various (exponential or larger) growth rates for the numbers of orbits on n-spaces. For example, if G fixes all parts of a partition of $\mathbf{Q}$ into r dense subsets, then the number of orbits on n-spaces is r^n.

Now Thomas' analogue of (3.10) is:

(5.19) *Let G be a collineation group of an infinite-dimensional projective space over $\mathrm{GF}(q)$, and suppose that G is transitive on n-spaces for all n.*

(i) If G is primitive on points, then its closure is the full collineation group of the projective space.

(ii) If G is imprimitive on points, then there is a congruence whose set of classes carries a G-invariant dense linear order.

There are two families of groups which are analogues of the imprimitive groups with $f_n = f_{n+1}$ for all even n (those with two congruence classes, or classes of size 2):

(a) The group preserving a symplectic form. The orbit of an n-space is determined by its rank, the codimension of its radical, which can be any even number not exceeding n; so there are $1 + \lfloor \frac{1}{2}n \rfloor$ orbits.

(b) Take a vector space of countable dimension over $\mathrm{GF}(q^2)$, and regard it as a $\mathrm{GF}(q)$-space by restricting scalars. Now our group is the group of $\mathrm{GF}(q^2)$-linear maps.

Thomas also characterised these examples, and proved partial analogues of the results described in §3.5. The results include the analogue of (3.15), and a theorem asserting that a group which is transitive on points and has equally many orbits on lines and planes (but is not transitive on planes) is contained in one of Examples (a) and (b) above (provided that $q > 2$).

It should be stressed that, if the results of Chapter 3 are incomplete, those here are even more so, and much remains to be discovered.

Exercise

Show that the group described in (b) above has $1 + \lfloor \frac{1}{2}n \rfloor$ orbits on n-subspaces.

Problem

What is the relationship between the "highly homogeneous" examples of Thomas and of Cameron and Hall?

5.7. ORBITS ON INFINITE SETS

What can be said about the orbits on infinite sets of a permutation group of countable degree?

First of all, any orbit on cofinite sets corresponds to the orbit on the complementary finite sets. We ignore these, and consider only *moieties*, infinite sets whose complements are also infinite.

We are no longer free to assume that our group is closed. For a countable highly transitive group has $2^{\aleph_0}$ orbits on moieties, but its closure is the symmetric group which has but one orbit. In a similar way, we have very different problems if we consider groups of degree greater than $\aleph_0$.

Two non-trivial results are known on this problem. But first, a triviality:

(**5.20**) *Let G be a permutation group with only finitely many orbits on moieties. Then G has a cofinite orbit, on which it acts highly transitively.*

For, by (3.3), $f_n(G)$ is bounded independent of n, and hence ultimately constant. then (3.15) shows that G has a cofinite orbit on which it acts highly homogeneously. But a highly homogeneous, not highly transitive group has infinitely many orbits on moieties (by inspection of the groups listed in (3.10)).

(**5.21**) *Let G be primitive but not highly transitive. Then G has $2^{\aleph_0}$ orbits on moieties.*

This was proved by Macpherson (1985a) at the same time, and with the same techniques, as his result (3.21) about growth rates. A similar encoding technique is used; instead of encoding exponentially many finite structures, he encodes $2^{\aleph_0}$ infinite ones.

The connection with growth rates is at least superficially plausible, in view of the description of ages as trees given in §4.10. This depends on the following result (see Exercise 4 of §2.6):

(**5.22**) *Let M be $\aleph_0$-categorical. Then any structure younger than M is embeddable in M.*

Now, if the growth rate of (f_n) (or, better, of (F_n)) is very fast, then we have many choices of infinite paths in the tree; such paths correspond to structures younger than M, and we'd expect many such structures, and hence many orbits of $\mathrm{Aut}(M)$

on moieties. But this idea seems to be too naïve to provide a proof. However, another, completely different, proof, related to the ideas of §5.3, has been given by Evans (1987a).

Finally, Neumann (1988) showed:

(5.23) *Suppose that G is primitive, and has no countable orbits on moieties. Then either G is highly transitive, or G acts on a projective or affine space over a finite field.*

To see this, note that G satisfies the condition (†) for a strongly minimal set in §5.3, that is, the stabiliser of a tuple has just one infinite orbit, which is cofinite. (For the stabiliser of a tuple has countable index in G, and hence cannot fix a moiety.) So the result follows, as does (5.12), from the classification of primitive groups with cofinite Jordan sets.

Here are a few examples. The non-trivial ones are due to Neumann.

1. The direct product $\mathrm{Sym}(n) \times S$ has $n + 1$ orbits on moieties, while $S \times S$ has countably many.

2. The automorphism group of a non-principal ultrafilter is highly transitive and has two orbits on moieties (the sets in the ultrafiler and their complements).

3. Let Λ be an ordered set. A permutation g of Λ is *piece-wise order-preserving* (pwop) if Λ is the union of finitely many subsets, such that the restriction of g to each subset is order-preserving. (This idea is due to Stoller (1963).) The group of pwop permutations of $\mathbf{N}$ is transitive on moieties; the group of pwop permutations of $\mathbf{Q}$ has $\aleph_1$ orbits on moieties. (The second fact depends on Laver's results (1973) on countable ordered sets.)

There has been some work on extending these results to higher cardinalities. See Shelah and Thomas (1988), Macpherson, Mekler and Shelah (to appear).

Exercises

1. If G has fewer than $2^{\aleph_0}$ orbits on moieties, then G is oligomorphic.

2. Verify Examples 2 and 3 above.

3. Show that the group of pwop permutations of any initial ordinal λ is transitive on moieties but has α^+ orbits on α-sets for any infinite cardinal $\alpha < \lambda$.

References

Adeleke, S. A. (1988), Embeddings of infinite permutation groups in sharp, highly transitive, and homogeneous groups, *Proc. Edinburgh Math. Soc.* (2) **31**, 169–178.

Adeleke, S. A. & Neumann, P. M. (to appear) (three papers).

Alper, T. M. (1987), A classification of all order preserving homeomorphisms of the reals that satisfy finite uniqueness, *J. Math. Psychol.* **31**, 135–154.

Anderson, R. D. (1958), The algebraic simplicity of certain groups of homeomorphisms, *Amer. J. Math.* **80**, 955–963.

Birch, B. J., Burns, R. G., Macdonald, S. O. & Neumann, P. M. (1976), On the degrees of permutation groups containing elements separating finite sets, *Bull. Austral. Math. Soc.* **14**, 7–10.

Brown, M. (1959), Weak n-homogeneity implies weak $n-1$-homogeneity, *Proc. Amer. Math. Soc.* **10**, 644–647.

Cameron, P. J. (1976), Transitivity of permutation groups on unordered sets, *Math. Z.* **148**, 127–139.

Cameron, P. J. (1978), Orbits of permutation groups on unordered sets, *J. London Math. Soc.* (2) **17**, 410–414.

Cameron, P. J. (1981), Orbits of permutation groups on unordered sets, II, *J. London Math. Soc.* (2) **23** 249–264.

Cameron, P. J. (1982), Colour schemes, *Ann. Discr. Math.* **15**, 81–95.

Cameron, P. J. (1983a), Orbits of permutation groups on unordered sets, III: imprimitive groups, *J. London Math. Soc.* (2) **27**, 229–237.

Cameron, P. J. (1983b), Orbits of permutation groups on unordered sets, IV: homogeneity and transitivity, *J. London Math. Soc.* (2) **27**, 238–247.

Cameron, P. J. (1987a), Portrait of a typical sum-free set, pp. 13–42 in *Surveys in Combinatorics* (ed. C. Whitehead), London Math. Soc. Lecture Notes **123**, Cambridge Univ. Press, Cambridge.

Cameron, P. J. (1987b), Some treelike objects, *Quart. J. Math. Oxford* (2), **38**, 155–183.

Cameron, P. J. (1987c), On the structure of a random sum-free set, *Probab. Th. Rel Fields* **76**, 523–531.

Cameron, P. J. (1989a), Some sequences of integers, *Discrete Math.* **75**, 85–102.

Cameron, P. J. (1989b), Groups of order-automorphisms of the rationals of prescribed scale type, *J. Math. Psychol.* **33**, 163–171.

Cameron, P. J. & Hall, J. I. (to appear), Some groups generated by transvection subgroups, *J. Algebra.*

Cameron, P. J. & Johnson, K. W. (1987), An essay on countable Burnside groups, *Math. Proc. Cambridge Philos. Soc.* **102**, 223–232.

Cameron, P. J. & Taylor, D. E. (1985), Stirling numbers and affine equivalence, *Ars Comb.* **20B**, 3–14.

Cameron, P. J. & Thomas, S. R. (1989), Groups acting on unordered sets, *Proc. London Math. Soc.* (3) **59**, 541–557.

Cantor, G. (1895), Beiträge zur Begründung der transfiniten Mengenlehre, *Math. Ann.* **46**, 481–512.

Chang, C. C. & Keisler, H. J. (1973), *Model Theory*, North-Holland, Amsterdam.

Cherlin, G., Harrington, L. & Lachlan, A. H. (1985), $\aleph_0$-categorical, $\aleph_0$-stable structures, *Ann. Pure Appl. Logic* **28**, 103–135.

Cohen, D. A. (1986), Thesis, Oxford University.

Comtet, L. (1974), *Advanced Combinatorics*, Reidel, Dordrecht.

Covington, J. (1986), Thesis, Oxford University.

Covington, J. (1989), A universal structure for N-free graphs, *Proc. London Math, Soc.* (3) **58**, 1–16.

Dixon, J. D., Neumann, P. M. & Thomas, S. R. (1986), Subgroups of small index in infinite symmetric groups, *Bull. London Math. Soc.* **18**, 580–586.

Droste, M. (1985), Structure of partially ordered sets with transitive automorphism groups, *Mem. Amer. Math. Soc.* **57**.

Droste, M., Holland, W. C. & Macpherson, H. D. (1989a), Automorphism groups of infinite semilinear orders, I, *Proc. London Math. Soc.* (3), **58**, 454–478.

Droste, M., Holland, W. C. & Macpherson, H. D. (1989b), Automorphism groups of infinite semilinear orders, II, *Proc. London Math. Soc.* (3), **58**, 479–494.

Droste, M. & Macpherson, H. D. (to appear), On k-homogeneous posets and graphs.

Ehrenfeucht, A. & Mostowski, A. (1956), Models of axiomatic theories admitting automorphisms, *Fund. Math.* **43**, 50–68.

Engeler, E. (1959), Äquivalenzklassen von n-Tupeln, *Z. Math. Logik Grundl. Math.* **5**, 340–345.

Erdős, P., Kleitman, D., & Rothschild, B. L. (1976), Asymptotic enumeration of K_n-free graphs, pp. 19–27 in *Teorie Combinatorie*, Accad. Naz. Lincei, Roma.

Erdős, P. & Rényi, A. (1963), Asymmetric graphs, *Acta Math. Acad. Sci. Hungar.* **14**, 295–315.

Evans, D. M. (1986a), Homogeneous geometries, *Proc. London Math. Soc.* (3) **52**, **305–327**.

Evans, D. M. (1986b), Subgroups of small index in infinite general linear groups, *Bull. London Math. Soc.* **18**. 587–590.

Evans, D. M. (1987a), Infinite permutation groups and minimal sets, *Quart. J. Math. Oxford* (2) **38**, 461–471.

Evans, D. M. (1987b), A note on automorphism groups of countably infinite structures, *Arch. Math. (Basel)* **49**, 479–483.

Evans, D. M. (to appear).

Evans, D. M. & Hewitt, P. R. (to appear), Counterexamples to a conjecture on relative categoricity.

Fraïssé, R. (1953), Sur certains relations qui généralisent l'ordre des nombres rationnels, *C. R. Acad. Sci. Paris* **237**, 540–542.

Fraïssé, R. (1986), *Theory of Relations*, North-Holland, Amsterdam.

Frasnay, C. (1974), Théorème de *G*-recollement (d'une famille d'ordres totaux). Application aux relations monomorphes, extension aux multirelations, pp 229–237 in *Permutations: Actes Colloq. Paris*, Gauthier-Villars, Paris.

Goulden, I. P. & Jackson, D. M. (1983), *Combinatorial Enumeration*, Wiley, New York.

Graham, R. L., Leeb, K., & Rothschild, B. L. (1972), Ramsey's theorem for a class of categories, *Adv. Math.* **8**, 417–433.

Graham, R. L., Rothschild, B. L. & Spencer, J. H. (1980), *Ramsey Theory*, Wiley, New York.

Hall, M. Jr. (1954), On a theorem of Jordan, *Pacific J. Math.* **4**, 219–226.

Hall, M. Jr. (1960), Automorphisms of Steiner triple systems, *IBM J, Res. Develop.* **4**, 460–472.

Harary, F. & Palmer, E. M. (1973), *Graphical Enumeration*, Academic Press, New York.

Hausdorff, F. (1914), *Grundzüge der Mengenlehre*, Leipzig.

Henson, C. W. (1971), A family of countable homogeneous graphs, *Pacific J. Math.* **38**, 69–83.

Higman, D. G. (1967), Intersection matrices for finite permutation groups, *J. Algebra* **6**, 22–42.

Higman, G. (1954), On infinite simple groups, *Publ. Math. Debrecen* **3**, 221-226.

Higman, G. (1977), Homogeneous relations, *Quart. J. Math. Oxford* (2) **28**, 31-39.

Hodges, W. A. (1989), Categoricity and permutation groups, pp. 53–72 in *Logic Colloquium '87* (ed. H.-D. Ebbinghaus *et al.*), Elsevier, Amsterdam.

Hodges, W. A., Lachlan, A. H. & Shelah, S. (1977), Possible orderings of an indiscernible sequence, *Bull. London Math. Soc.* **9**, 212–215.

Hodkinson, I. M. & Macpherson, H. D. (1988), Relational structures determined by their finite substructures, *J. Symbolic Logic* **53**, 222–230.

Hrushovski, E. (to appear).

Huntingdon, E. V. (1904), The continuum as a type of order: an exposition of the modern theory, *Ann. Math.* **6**, 178–179.

Joyal, A. (1981), Une theorie combinatoire des séries formelles, *Adv. Math* **42**, 1–82.

Kantor, W. M. (1972), On incidence matrices of projective and affine spaces, *Math. Z.* **124**, 315–318.

Kantor, W. M., Liebeck, M. W. & Macpherson, H. D. (1989), $\aleph_0$-categorical structures smoothly approximable by finite substructures, *Proc. London Math. Soc.* (3) **59**, 439–463.

Kueker, D. W. (1968), Definability, automorphisms and infinitary languages, pp. 152–165 in *The Syntax and Semantics of Infinitary Languages* (ed. J. Barwise), Lecture Notes in Math. 72, Springer, Berlin.

Lachlan, A. H. (1974), Two conjectures on the stability of ω-categorical theories, *Fund. Math.* **81**, 133–145.

Lachlan, A. H. (1984), Countable homogeneous tournaments, *Trans. Amer. Math. Soc.* **284**, 431–461.

Lachlan, A. H. & Woodrow, R. E. (1980), Countable ultrahomogeneous undirected graphs, *Trans. Amer. Math. Soc.* **262**, 51–94.

Lascar, D. (1982), On the category of models of a complete theory, *J. Symbolic Logic* **47**, 249–266.

Laver, R. (1973), An order type decomposition theorem, *Ann. Math.* (2) **98**, 96–119.

Livingstone, D. & Wagner, A. (1965), Transitivity of finite permutation groups on unordered sets, *Math. Z.* **90**, 393–403.

Macpherson, H. D. (1983), The action of an infinite permutation group on the unordered subsets of a set, *Proc. London Math. Soc.* (3) **46**, 471–486.

Macpherson, H. D. (1985a), Orbits of infinite permutation groups, *Proc. London Math. Soc.* (3) **51**, 246–284.

Macpherson, H. D. (1985b), Growth rates in infinite graphs and permutation groups, *Proc. London Math. Soc.* (3) (**51**, 285–294.

Macpherson, H. D. (1986a) Groups of automorphisms of $\aleph_0$-categorical structures, *Quart. J. Math. Oxford* (2) **37**, 449–465.

Macpherson, H. D. (1986b), Homogeneity in infinite permutation groups, *Period. Math. Hungar.* **17**, 211–233.

Macpherson, H. D. (1987), Permutation groups of rapid growth, *J. London Math. Soc.* (2) **35**, 276–286.

Macpherson, H. D., Mekler, A. H., and Shelah, S. (to appear).

Macpherson, H. D. & Neumann, P. M. (to appear), Subgroups of infinite symmetric groups.

Macpherson, H. D. & Praeger, C. E. (to appear), Maximal subgroups of infinite symmetric groups.

Macpherson, H. D. & Thomas, S. R. (to appear), Reducts of the random graph.

McCleary, S. H. (1985), Free lattice-ordered groups represented as o-2-transitive *l*-permutation groups, *Trans. Amer. Math. Soc.* **290**, 69–79.

McDonough, T. P. (1977), A permutation representation representation of a free group, *Quart. J. Math. Oxford* (2) **28**, 353–356.

Mekler, A. H. (1986), Groups embeddable in the autohomeomorphisms of **Q**, *J. London Math. Soc.* (2) **33**, 49–58.

Miller, C. C. (to appear), Imprimitive automorphism groups.

Morley, M. (1965), Categoricity in power, *Trans. Amer. Math. Soc.* **114**, 514–538.

Mura, R. B., and Rhemtulla, A., (1977), *Orderable Groups*, Marcel Dekker, New York.

Neumann, B. H. (1949), On orderable groups, *Amer. J. Math.* **71**, 1–18.

Neumann, B. H. (1954), Groups covered by permutable subsets, *J. London Math. Soc.* **29**, 236–248.

Neumann, P. M. (1975), The lawlessness of finitary permutation groups, *Arch. Math. (Basel)*, **26**, 561–566.

Neumann, P. M. (1976), The structure of finitary permutation groups, *Arch. Math. (Basel)*, **27**, 3–17.

Neumann, P. M. (1985a), Automorphisms of the rational world, *J. London Math. Soc.* (2) **32**, 439–448.

Neumann, P. M. (1985b), Some primitive permutation groups, *Proc. London Math. Soc.* (3) **50**, 265–281.

Neumann, P. M. (1988), Homogeneity of infinite permutation groups, *Bull. London Math. Soc.* **20**, 305–312.

Oxtoby, J. C. (1980), *Measure and Category*, Springer, Berlin.

Pillay, A. (1983), *An Introduction to Stability Theory*, Oxford Univ. Press, Oxford.

Pouzet, M. (1976), Application d'une propriété combinatoire des parties d'un ensemble aux groupes et aux relations, *Math. Z.* **150**, 117–134.

Pouzet, M. (1981), Application de la notion de relation presque-enchaînable au dénombrement des restrictions finies d'une relation, *Z. Math. Logik Grundl. Math.* **27**, 289–332.

Rota, G.-C. (1964), On the foundations of combinatorial theory, I: Theory of Möbius functions, *Z. Wahrsch. verw. Geb.* **2**, 340–368.

Rubin, M. (to appear), On the reconstruction of $\aleph_0$-categorical structures from their automorphism groups.

Ryll-Nardzewski, C. (1959), On category in power $\leq \aleph_0$, *Bull. Acad. Pol. Sci. Sér. Math. Astr. Phys.* **7**, 545–548.

Schreier, J. & Ulam, S. M. (1933), Über die Permutationsgruppe der natürlichen Zahlenfolge, *Studia Math.* **4**, 134–141.

Seidel, J. J. (1976), A survey of two-graphs, p. 481–511 in *Teorie Combinatorie*, Accad. Naz. Lincei, Roma.

Semmes, S. W. (1981), Endomorphisms of infinite symmetric groups, *Abstracts Amer. Math. Soc.* **2**, 426.

Shelah, S. (1978), *Classification Theory and the Number of Non-isomorphic Models*, North-Holland, Amsterdam.

Shelah, S. & Thomas, S. R. (1988), Implausible subgroups of infinite symmetric groups, *Bull. London Math. Soc.* **20**, 313–318.

Shelah, S. & Thomas, S. R. (1989), Subgroups of small index in infinite symmetric groups, *J. Symbolic Logic* **54**, 95–99.

Sierpiński, W. (1920), Une propriété topologique des ensembles dénombrables denses en soi, *Fund. Math.* **1**, 11–16.

Skolem, S. (1920), Logisch-kombinatorische Untersuchung, *Skrifer uitgit av Videnskap*, I Kl., Kristiania.

Stevens, S. S. (1946), On the theory of scales of measurement, *Science* **103**, 677–680.

Stoller, G. (1963), Example of a proper subgroup of S_∞ which has a set-transitivity property, *Bull. Amer. Math. Soc.* **69**, 220–221.

Svenonius, L. (1959), $\aleph_0$-categoricity in first-order predicate calculus, *Theoria* **25**, 82–94.

Thomas, S. R. (1986), Groups acting on infinite-dimensional projective spaces, *J. London Math. Soc.* (2) **34**, 265–273.

Thomas, S. R. (1988), Groups acting on infinite-dimensional projective spaces II, *Proc. London Math. Soc.* (3), **56**, 511–528.

Tits, J. (1952), Généralisation des groupes projectifs basée sur leurs propriétés de transitivité, *Acad. Roy. Belgique Cl. Sci. Mem.* **27**.

Tits, J. (1974), *Buildings of Spherical Type and Finite BN-Pairs*, Lecture Notes in Math. 382, Springer, Berlin.

Truss, J. K. (1985), The group of the countable universal graph, *Math. Proc. Cambridge Philos. Soc.* **98**, 213–245.

Truss, J. K. (1986), Embeddings of infinite permutation groups, pp. 335–351 in *Groups – St. Andrews 1985* (ed. E. F. Robertson and C. M. Campbell), *London Math. Soc. Lecture Notes* **121**, Cambridge Univ. Press, Cambridge.

Truss, J. K. (1989a), Infinite permutation groups, I: Products of conjugacy classes, *J. Algebra* **120**, 454–493.

Truss, J. K. (1989b), Infinite permutation groups, II: subgroups of small index, *J. Algebra* **120**, 494–515.

Truss, J. K. (1989c), The group of almost automorphisms of the countable universal graph, *Math. Proc. Cambridge Philos. Soc.* **105**, 223–236.

Truss, J. K. (to appear), Infinite simple permutation groups — a survey.

Vaught, R. L. (1963), Models of complete theories, *Bull. Amer. Math. Soc.* **69**, 299–313.

White, S. (1988), The group generated by $x \mapsto x + 1$ and $x \mapsto x^p$ is free, *J. Algebra* **118**, 408–422.

Wiegand, R. (1978), Homeomorphisms of affine surfaces over a finite field, *J. London Math. Soc.* (2) **18**, 28–32.

Wielandt, H. (1959), *Unendliche Permutationsgruppen*, Universität Tubingen.

Wielandt, H. (1964), *Finite Permutation Groups*, Academic Press, New York.

Wright, E. M. (1980), The number of connected sparsely-edged graphs, *J. Graph Theory* **4**, 393–407.

Yoshizawa, M. (1979), On infinite four-transitive permutation groups, *J. London Math, Soc.* (2) **19**, 437–438.

Zil'ber, B. I. (1979), Totally categorical theories: structural properties and non-finite axiomatisability, pp. 381–410 in *Lecture Notes in Math.* **834**, Springer, Berlin.

Zil'ber, B. I. (1984), The structure of models of uncountably categorical theories, pp. 359–368 in *Proc. Int. Congr. Math.* (Warsaw 1983), Warsaw.

Zil'ber, B. I. (1988), Finite homogeneous geometries, pp. 186-208 in *Proc. 6th Easter Conf. Model Theory* (ed. B. Dahn & H. Wolter), Humboldt Univ., Berlin.

Index

A

$\aleph_0$-categorical, 5, 30, *passim*
abelian group, 90, 97, 130
abelian-by-finite group, 115
absolute zero, 101
action, 6, 21, 59, 124
Adeleke, S. A., 3, 104, 105, 121–124, 128
affine group, 60, 66, 101, 119–121, 129
age, 32, 33, 111–116, 135
algebra, 78
algebraic closure, 36, 107, 109, 127, 131
algebraic geometry, 137
algebraically closed field, 119
almost automorphism, 47, 110
almost homogeneous, 93, 96
almost transitive, 87
Alper, T. L., 102
alternating group, 4, 38
amalgamation property, 32
Anderson, R. D., 4
angle, 58
anti-automorphism, 45, 58
Axiom of Choice, 46, 109
axioms for a group, 6, 10

B

B-group, 90
back and forth, 31, 40, 124, 126, 138
Baire category theorem, 13, 29, 103
Bell numbers, 23
betweenness, 61, 71, 129
betweenness, generalised, 122
Bezout's theorem, 137
binary tree, 77
bipartite graph, 72, 93, 114
Birch, B. J., 37, 38
black, 134
blue, 64, 126

Bolzano-Weierstrass theorem, 18
Boolean algebra, 108
boron tree, 34, 35, 58, 66, 71, 124, 134
boron-carbon tree, 66
Brown, M. 4
Burns, R. G., 37, 38
Burnside, W., 90

C

canonical relational structure, 26, 28, 33, 80
Cantor, G., 124
Cantor ternary set, 14
Cantor-Bendixson rank, 130
Cantor's theorem, 30, 124, 129
Cartesian product, 7, 8
Catalan numbers, 59
category, 5, 13, *passim*
chain relation, 123, 129, 134
chainable, 62
Chang, C. C., 5
Cherlin, G. L., 108, 120, 130, 132
Chinese Remainder Theorem, 46
closure, 28
cofinitary, 99, 104–106
Cohen, D. A., 102
colouring, 16, 53, 64, 67
colour scheme, 64, 67
commutative diagram, 8, 32
Compactness Theorem, 11, 18, 62, 130
complement (of hypergraph), 36, 45, 135
complete metric space, 13, 28
complete theory, 12
Completeness Theorem, 11, 39, 40
Comtet, L., 58, 78
congruence, 8, 107, 141
connectedness, 79
connectives, 10

Printed in the United States
By Bookmasters